Artificial intelligence in application

Thomas Barton · Christian Müller

Editors

Artificial intelligence in application

Legal aspects, application potentials and use scenarios

 Springer

Editors
Thomas Barton
Department of Computer Science
Worms University of Applied Sciences
Worms, Germany

Christian Müller
Faculty of Business, Computing and Law
Technical University of Applied
Sciences Wildau
Wildau, Germany

ISBN 978-3-658-43842-5   ISBN 978-3-658-43843-2   (eBook)
https://doi.org/10.1007/978-3-658-43843-2

This book is a translation of the original German edition "Künstliche Intelligenz in der Anwendung" by Thomas Barton, published by Springer Fachmedien Wiesbaden GmbH in 2021. The translation was done with the help of an artificial intelligence machine translation tool. A subsequent human revision was done primarily in terms of content, so that the book will read stylistically differently from a conventional translation. Springer Nature works continuously to further the development of tools for the production of books and on the related technologies to support the authors.

This Springer imprint is published by the registered company Springer Fachmedien Wiesbaden GmbH, part of Springer Nature.
The registered company address is: Abraham-Lincoln-Str. 46, 65189 Wiesbaden, Germany

If disposing of this product, please recycle the paper.

Contents

List of Contributors

Thomas Barton Department of Computer Science, Worms University of Applied Sciences, Worms, Germany

Stephan Böhm Hochschule RheinMain, Wiesbaden, Germany

Christian Czarnecki Hochschule Hamm-Lippstadt, Hamm, Germany

Michael Gröschel Hochschule Mannheim, Mannheim, Germany

Gerhard Hausmann Infrastruktur- und Architekturmanagement, Barmenia, Wuppertal, Germany

Marcel Herold Technische Hochschule Wildau, Wildau, Germany

Georg Rainer Hofmann Information Management Institut, Aschaffenburg, Germany

Wolfgang Jäger Dr. Jäger Management Group, Königstein im Taunus, Germany

Uwe Lämmel Hochschule Wismar, Wismar, Germany

Olena Linnyk Milch & Zucker AG, Giessen, Germany

Stephan Meyer Technische Hochschule Wildau, Wildau, Germany

Christian Müller Faculty of Business, Computing and Law, Technical University of Applied Sciences Wildau, Wildau, Germany

Carl-Christian Neundorf aubex GmbH, Hockenheim, Germany

Nikolai Nölle , Bornheim, Germany

Salmai Qari Hochschule für Wirtschaft und Recht, Berlin, Germany

Marc Roedenbeck Technische Hochchule Wildau, Wildau, Germany

Gabriele Roth-Dietrich Hochschule Mannheim, Mannheim, Germany

Dominik Schneider , Bad Honnef, Germany

Meike Schumacher Information Management Institute, Aschaffenburg, Germany

Ingolf Teetz Milch & Zucker AG, Giessen, Germany

Mathias Walther Technische Hochschule Wildau, Wildau, Germany

Frank Wisselink , Bonn, Germany

Part I

Introduction

Potential for Artificial Intelligence and Its Application

1

Thomas Barton and Christian Müller

What is artificial intelligence? "The study of how to make computers do things at which, at the moment, people are better" (Elaine Rich and Kevin Knight, 1991)

Abstract

This chapter gives an introduction into the book "Artificial intelligence in application". It is a translated and updated version of the book "Data Science anwenden", which was published in German language in origin. Within this chapter, all contributions of the whole book are presented in three thematic focal points starting with the topic legal aspects. Most of the contributions are focusing on the description of application potential for artificial intelligence and on the presentation of concrete application scenarios.

This book focuses on the application of artificial intelligence. The description of legal aspects forms the basis of this volume. The description of different application potential and the presentation of concrete application scenarios are the focus of this volume.

T. Barton (✉)
Department of Computer Science, Worms University of Applied Sciences, Worms, Germany
e-mail: barton@hs-worms.de; https://www.prof-barton.de

C. Müller
Faculty of Business, Computing and Law, Technical University of Applied Sciences Wildau, Wildau, Germany
e-mail: christian.mueller@th-wildau.de; https://www.th-wildau.de/christian-mueller/

1.1   Legal Aspects

Based on the European General Data Protection Regulation, legal aspects of automated decision-making are highlighted and challenges of artificial intelligence for our legal system are explained.

In his contribution (Chap. 2) "Legal Challenges of Artificial Intelligence and How to manage Them", Stephan Meyer shows that automated decisions, which are enabled by Artificial Intelligence, pose challenges for the legal system. In order to exclude gaps in legal accountability, specific regulations for different areas of use are necessary. He demonstrates this with reference to autonomous driving.

1.2   Potential Applications

The application potential of artificial intelligence is examined by examples in individual industries and applications in a classical as well as in a disruptive environment. Within the framework of this thematic volume five teams of authors have come together, consisting of a single person or up to four authors.

The article (Chap. 3) "Application Potential of Artificial Intelligence in the Car Trade" by Michael Gröschel, Gabriele Roth-Dietrich and Carl-Christian Neundorf uses four mobility scenarios to examine how important business processes in car retailing could change in the future. These processes are car sales, maintenance, repair, and purchasing of used cars. They identify use cases for the application of various artificial intelligence methods and present them in their paper.

Georg Rainer Hofmann and Meike Schumacher examine the acceptance of artificial intelligence methods in retailing in their article (Chap. 4) entitled "Acceptance of AI systems in Retail". They focus on five application scenarios. These scenarios are evaluated with the help of expert interviews.

The contribution (Chap. 5) "Application Potential for Causal Inference in Online Marketing" by Thomas Barton is dedicated to the joint application of causal diagrams and statistical analyses in the subject area of online marketing. He turns to the question how to assess and optimize the effectiveness of online campaigns.

Dominik Schneider, Frank Wisselink, Nikolai Nölle and Christian Czarnecki shed light on the potential of intelligent services in their article (Chap. 6) "Influence of Artificial Intelligence on Customer Journeys Using the Example of Intelligent Parking". They show by means of smart parking, how services will influence customers when making commercial decisions.

The article (Chap. 7) by Gabriele Roth-Dietrich is devoted to the influence of artificial intelligence on platform business models. She examines the significance of Artificial Intelligence for the design of platform companies and derives recommendations for action.

1.3   Operational Scenarios

Concrete operational scenarios for Artificial Intelligence are presented using four examples. These scenarios cover a wide range of applications. They include the automation of document processing, the analysis of application letters on the one hand and service interviews on the other as well as the formulation of job advertisements.

In their article (Chap. 8), Gerhard Hausmann and Uwe Lämmel present an operational scenario for digitizing document processing with the help of artificial intelligence. Their article entitled "Artificial Intelligence in Automated Document Processing Using the Example of Health Insurance Companies" shows how invoices from a private health insurance company are processed automatically.

Two contributions are dedicated to the topic of human resource management. In their contribution, Stephan Böhm, Olena Linnyk, Wolfgang Jäger and Ingolf Teetz present the status of development and introduce concrete application examples. Their contribution (Chap. 9) entitled "AI in Recruiting: Potentials, Status Quo, and Pilot Projects in Germany" also presents findings from concrete pilot applications. Marc R.H. Roedenbeck, Salmai Qari, and Marcel Herold analyze job application letters in their contribution (Chap. 10). Their contribution entitled "Artificial Intelligence in the Staffing Process: Performance Comparisons of (Un)supervised Learning for the Screening of Job Applications" uses methods such as cluster analysis or a neural network within the framework of a pilot study.

Finally, Mathias Walther dedicates his contribution (Chap. 11) to the topic of call centers and presents a framework for the analysis of conversations. His contribution is entitled "An AI-based Framework for Speech and Voice Analytics to Automatically Assess the Quality of Service Conversations".

Thomas Barton studied physics at the TU Kaiserslautern with minors in computer science and mathematics. He obtained his doctorate at the intersection of physics and medicine in the research group of Prof. Dr. Wolfgang Demtröder in the Department of Physics at TU Kaiserslautern. During this work, he gained relevant experience in the software-supported control of experiments and in the analysis of large amounts of data. This expertise led him to SAP SE, where he worked for 10 years with a focus on application development, also consulting, training and project management. Since 2006, he has been working at Worms University of Applied Sciences as a professor of computer science with a focus on business informatics. His work focuses on the development of business applications, e-business and data science. He is the author and editor of numerous publications. In addition, he is active in various committees and expert panels. He is also the spokesperson for the GI Advisory Board for Universities of Applied Sciences.

Prof. Dr. Christian Müller studied mathematics at the Free University of Berlin and received his PhD in 1989 on network flows with constraints. From 1990 to 1992 he worked at Schering AG and from 1992 to 1994 at Berliner Verkehrsbetriebe (BVG) in the area of timetable and duty schedule optimization. In 1994 he received an appointment at the Technical University of Wildau, Germany. His main research interests are the design of information systems, mathematical optimization and the simulation of business processes.

Part II

Legal Aspects

Legal Challenges of Artificial Intelligence and How to Manage Them

2

Stephan Meyer

Abstract

Artificial intelligence performs complex assessments as its basis for making decisions or prompting technical processes without any human involvement. Such autonomous entities pose challenges to the legal system. They are unfit as bearers of legal responsibilites and rights. As a result, they leave gaps in legal accountability when they replace humans. Legislators will close these gaps in an innovation-friendly way only if regulation of artificial intelligence will be specific to each individual field of application. This has been achieved to some extent for autonomous driving.

2.1 Introduction

Artificial intelligence (AI) systems are often labeled "autonomous" systems. The predicate of autonomy signifies those systems' ability to generate a problem-solving algorithm by themselves (in the case of machine learning (ML)), or to arrive at original solutions for complex tasks in other ways. Thus, an autonomous system does more than just to select—based on input data—an answer predetermined by humans and to refine it at best (*autonomous* vs. *automatic*). This is reflected by the common definitional approach that AI handles tasks that previously would have required human thinking.

S. Meyer (✉)
Technische Hochschule Wildau, Wildau, Germany
e-mail: stephan.meyer@th-wildau.de

Machines that make decisions "freely"[1] within their field of application give rise to unfamiliar legal questions. Striking examples from civil, criminal and public law illustrate both the challenges posed by machine autonomy (as opposed to mere automation) and the existing strategies for managing them (Sect. 2.2). They are evidence of a need for regulation. That need has been recognized by European and German authorities just as much as the technology's tremendous impact on competition. Section 2.3 therefore defines general requirements for innovation-friendly regulation, evaluates the regulatory guidelines of the Federal Government's Data Ethics Commission, and takes a brief look at the most recent EU Artificial Intelligence Act (AIA).

2.2   Machine Autonomy as a Legal Challenge: Examples

2.2.1   AI and Copyright Law

In several Directives, European Union law requires an "own intellectual creation"[2] for a work to be covered by copyright protection. The European Court of Justice (ECJ) considers this requirement to be a principle of Union law. According to German law, works need to be "personal intellectual creations" in order to be protected, Sect. 2(2) Act on Copyright and Related Rights (UrhG).[3,4] Thus, creative activity of a human being is required. Sect. 11 first sentence UrhG confirms this by stating that it is the author's intellectual and personal relationships to the work that is protected by copyright. As a result, mere artistic appearance of an object is not enough to be a protected "work" in the legal sense.

The use of tools occurs almost always in the course of the creation of a work; just think of the painter's brush. That is unproblematic as long as there still exists a significant mental effort on the part of the creator that possesses relevant influence on the final product. Therefore, a work worthy of protection can exist even if individual components were created automatically.[5]

---

[1] The term is used here in a general sense and is not intended to invoke all philosophical baggage that comes with it.

[2] Art. 3 para. 1 first sentence Directive 96/9/EC of 11 March 1996 on the legal protection of databases (OJ L 77, 27 March 1996, p. 20), as amended by Directive (EU) 2019/790 (OJ L 130, 17 May 2019, p. 92); Art. 6 first sentence Directive 2006/116/EC of 12 Dec. 2006 on the term of protection of copyright and certain related rights (OJ. L 372, 27 Dec. 2006, p. 12), as last amended by Directive 2011/77/EU (OJ L 265, 11 Oct. 2011, p. 1); Article 1(3) first sentence Directive 2009/24/EC of 23 April 2009 on the legal protection of computer programs (OJ L 111, 5 May 2009, p. 16); see also [16, p. 1255 et seq.].

[3] ECJ, Judg. of 16 June 2009, Case C-5/08—Infopaq International A/S/Danske Dagblades Forening, para. 36; see also [16, 46, 61].

[4] On the unclear relationship between the European and the German legal concept of a creative "work" [8, marginal no. 13–14.].

[5] See also ECJ, Judg. of 1 Dec. 2011, Case C-145/10—Eva-Maria Painer v Standard VerlagsGmbH and others, para. 88–89.

More recently, however, "artistic" and "literary" works created solely by an artificial intelligence have attracted widespread attention. These works include "The Next Rembrandt", a "new" Rembrandt created by a self-learning AI based on an approximation of the painter's style. A collection of Rembrandt paintings served as training data.[6] In contrast, the AI-generated painting "Edmond de Belamy"—auctioned for $432,000—is based on an analysis of 15,000 paintings created between the fourteenth and nineteenth century.[7]

An assessment of the protectability of such AI "works" is facilitated by an older problem, that is, art created by a random generator. Such art, too, requires a certain level of human influence on the final product in order to be a copyright-protected "work". For example, the author could create some basic patterns, which the random generator then arranges [32, pp. 971–72]. However, the activity of the random generator itself only presents an imitation of those basic patterns, i.e., of the artist's *style*. A mere style as such is not eligible for protection, and the specific creative influence of the human being on the final product is missing. Therefore, an additional requirement for work worthy of protection is suggested: The creator may need to *select* which of the various outputs of the random generator is actually to be presented as a work of art [32, p. 972, 69, para. 8, para 17]. Others, however, contest with good reasons that merely finding and presenting an object that appears "artistic" could prompt copyright protection [1, marginal no. 54–55, 35, p. 577, 8, p. 247, 47, p. 808, 60, marginal no. 25–26].

For AI "works" it follows that copyright protection at least requires that the training data are based on creations by the human artist employing the AI. The above cases do not even meet this basic requirement.[8] In those cases that do, protection would furthermore depend on accepting the contested "selection theory".[9]

To conclude, copyright protection for AI-generated "works" does rather not exist as the law stands at present. Whether AI "works" should at least be covered by a newly to be introduced ancillary copyright is intensely discussed in the legal literature [16, p. 1257 et seq., 32, p. 973 et seq., 35, p. 577, 47, p. 809 et seq.]. The Legal Affairs Committee of the European Parliament even called for establishing full-fledged copyright protection [23, p. 33]. Such considerations are to be appreciated, for it is obvious that "creative" AI will become increasingly more intelligent and will therefore transcend mere eclecticism—for example, by acquiring a sense of aesthetics in connection with illustrative material from

---

[6] https://www.vml.com/work/next-rembrandt (03/26/2022).

[7] https://obvious-art.com/portfolio/edmond-de-belamy/; https://time.com/5357221/obvious-artificial-intelligence-art/ (03/26/22).

[8] Coming to the same conclusion: [32, 47].

[9] In contrast, it does not appear to be problematic if a creator of a work only has individual aspects or excerpts designed by AI; in this case, a significant creative influence of the human creator on the final product is maintained, cf. [83, para. 1].

everyday life.[10] Without protection, however, there is no incentive to create. The legislator has considerable leeway in this area to allocate rights under innovation policy aspects.

2.2.2 Autonomous Driving

Technical systems for vehicle control are commonly classified into five levels based on the extent to which they take over driving functions (SAE J3016 Standard).[11] They range from traditional assistance systems such as cruise control at the first level via three levels of automation to "autonomous driving" in the narrower sense at the highest level, where humans on board will only be passengers. Automated/autonomous driving systems also use AI/ML.

A (partially) self-driving vehicle raises a number of questions that reflect the faltering of established schemes of legal attribution. However, in contrast to copyright law, it is not primarily a question of owning rights, but of having obligations in road traffic—and thus of the allocation of responsibility.

Presently, persons responsible under German road traffic law have been in particular the operator and the keeper of the vehicle. There is no legal definition of the term "operator of the vehicle" ("Fahrzeugführer"). Nevertheless, the official title of Sect. 31 of the Road Traffic Licensing Regulation (StVZO) ("Responsibility for the operation of vehicles") in conjunction with Sect. 1 of the regulation, which requires the operator to be qualified to "independently oversee" vehicle operation, indicate that the "operator" is responsible for "overseeing" the operation of the vehicle. A further indication of the legal role of the operator is provided by Sect. 23 of the Traffic Code (StVO). Its official title (*Other* Duties of Vehicle Operators) implies that the duties laid down earlier in Sect. 1 through 22 StVO concerning traffic with vehicles also address the operator of the vehicle (even though there is no express reference to an operator of the vehicle in these Sections). The operator is therefore responsible for compliance with traffic regulations when operating the vehicle. Consequently, court rulings deem "the person who uses all or at least some of the essential technical equipment of the vehicle intended for its movement" to legally be "the operator". "Thus, in order to be the operator of a vehicle, a person must, in sole or joint responsibility, set the vehicle in motion by using its propulsive power in accordance with its intended purpose, or steer the vehicle wholly or in part while it is moving in public traffic by using its technical equipment".[12]

In contrast, the vehicle keeper's responsibility results from his or her creating an exceptional safety hazard by keeping available a motor vehicle [80, Sect. 7 marginal no. 3]. Sect. 7 of the Road Traffic Act (StVG) accommodates this hazard by establishing strict liability of the keeper.

---

[10] Optimism also at [16, pp. 1254–55].

[11] https://www.sae.org/standards/content/j3016_201806/ (03/26/2022).

[12] BGHSt [= German Federal Court of Justice, Criminal Cases Reports] 35, 390 (393–94).

2.2.2.1 Legality of the Use of Driving Systems: The StVG Amendment

With road traffic law establishing the driver's responsibility for compliance with traffic rules while driving (see above), permissibility of partially transferring vehicle control to a machine seems barely conceivable—at least if the transfer were to include responsibility for compliance with traffic rules (counter-example: a simple cruise control). The legislator recognized this obstacle to innovation and, in 2017, added Sect. 1a, 1b and 63a to the Road Traffic Act—among other amendments.[13] The core provisions of the amendment being of interest here are:

1. Admissibility of motor vehicle operation by means of highly or fully automated driving functions (SAE levels 3 and 4) when used as intended (Sect. 1a para. 1 StVG)
2. Requirement for such a driving function to comply with the traffic rules directed at the authority that is overseeing vehicle operation while the vehicle is being driven (Sect. 1a para. 2 no. 2 StVG)
3. Ability of the driving function to indicate to the driver in sufficient time a need for his or her resuming vehicle control (Sect. 1a para. 2 no. 5 StVG)
4. The user of the driving function remains the operator of the vehicle in the legal sense even during automated operation (Sect. 1a para. 4 StVG).
5. The operator may turn away from the traffic situation and from vehicle control during automated operation *(authorization to turn away);* when doing so, he or she must remain perceptive in such a way that he or she can resume vehicle control immediately if the automated system prompts him or her to do so, or if resuming control is required by compelling circumstances (Sect. 1b StVG).
6. Ability of the vehicle to store position and time when there is a change in vehicle control between operator and automated system (Sect. 63a StVG).

With this amendment the Federal Government aims to establish a German pioneering role in automated and connected driving and to ensure legal certainty for the use of driving systems.[14] The Coalition Agreement between the majority political parties for the 19th legislative period of the German Bundestag [= Federal Parliament] provided that the next step should be to adopt another amendment of the StVG that sets the stage for fully autonomous vehicles (SAE Level 5).[15] However, the most recent "Autonomous Driving Act" that entered into force in July 2021 still does not go beyond Level 4.[16] While the Act now refers to "autonomous driving", and not just to "fully automated driving" (comp. core provision no. 1 above), the use of the autonomous function is limited to a specified scope of operation. The official explanatory memorandum for the draft bill expressly mentions this deviation from the earlier coalition agreement.

---

[13] Art. 1 No. 1 and 5 of the Eighth Act amending the Road Traffic Act of 16 June 2017 (BGBl [= Federal Law Gazette] I 1648).

[14] Bundestag [German Federal Parliament] printed matter 18/11300 of 20 Feb. 2017, p. 13.

[15] Coalition Agreement between CDU, CSU and SPD, 19th legislative period, p. 80.

[16] Bundestag [German Federal Parliament] printed matter 14 Feb. 2020, p. 10.

Aside from some individual criticism,[17] the amendments justify optimism. The legislator recognises the innovation potential of AI in the traffic sector and has therefore invoked the enabling function of law for new technologies. The new rules for automated/autonomous driving functions allow vehicle operators for the first time to turn away from the traffic situation without failing to exercise proper care.[18] This means that the operator can rely on the system[19] and is thus spared from being prosecuted in the event of irregular driving.[20] Likewise, negligent liability from Sect. 18 para. 1 StVG (limited in amount, Sect. 12 StVG) and from Sect. 823 et seq. German Civil Code (BGB) (not limited in amount) does not apply. Nevertheless, there is commonly no gap in liability for victims of traffic accidents: The vehicle keeper's strict liability (Sect. 7 StVG, see Sect. 2.2.2 above) will step in. All these provisions of the StVG amendments obviously favour the technology's chances of market penetration.

The legislator should also be commended for demonstrating great care as to the wording of the amendments. Sect. 1a para. 2 no. 2 StVG states that the automated driving function needs to be able to comply with the traffic rules directed "at the authority that is overseeing vehicle operation" ("Fahrzeugführung") (see core provision no. 2 above). This wording reflects the non-human nature of the agent that controls vehicle operation. While the user of the automated driving function remains the operator of the vehicle in the legal sense (Sect. 1a para. 4 StVG, see core provision no. 4 above), he or she is, however, no longer required to personally ensure compliance with traffic regulations ("authorization to turn away", see core provision no. 5 above). Rather, this is the automated driving function's job. For this reason, the legislator distinguishes between the (human) "operator" and the "authority that is overseeing vehicle operation" (the latter including the automated driving function). The term "authority that is overseeing vehicle operation" is used in Sect. 1a para. 2 no. 2 StVG only and thus exclusively in the context of automated/autonomous driving. The legislator requires this authority to "comply" with traffic regulations—and not to follow or to observe them. The latter two terms (which are otherwise common in German law) would imply imposing an "obligation". The notion of "obligation" is meaningful only to entities with free will. Therefore, no "obligations" can be addressed to machines. Consequently, the legislator chose a neutral wording [54, p. 236].

---

[17] For instance, at [48, 59, 63].

[18] The driver retains the legal quality of being the "operator of the vehicle" while using the automated driving function, see previously in the text (core provision no. 4). He or she remains the addressee of traffic rules and therefore needs proof in the event of a traffic violation that he or she exercised appropriate care in accordance with core provision no. 5. Data storage pursuant to Sect. 63a StVG (core provision no. 6) serves that purpose.

[19] Bundestag [German Federal Parliament] printed matter 18/11300 of 20 Feb. 2017, pp. 13–14.

[20] Cf. [66, pp. 360–61].

2.2.2.2 Unpredictability of Autonomous Systems as a Barrier to Innovation?

2.2.2.2.1 Unpredictability as an Inherent Quality of Autonomous Systems

However, a number of autonomy-specific challenges remain unaddressed and will require timely attention from the legislator.

If a technical system has the ability to generate and improve problem-solving algorithms, it can also arrive at unexpected or undesired answers (often dubbed "unpredictability"). The problem-solving algorithm, once generated, cannot be conclusively examined ex-ante for sources of possible misevaluations in future practical use due to its complexity. Likewise, future misevaluations cannot be completely ruled out with the help of practical test runs [55, 65], cf. [78, pp. 477–78]. Thus, the "unpredictability" of autonomous systems establishes a new type of hazard that is inherently different from a traditional technical malfunction or failure: The "autonomy risk".[21]

2.2.2.3 Legal Issues of Unpredictability

2.2.2.3.1 Liability Law

In liability law, the autonomy risk brings another party bearing responsibility more to the fore[22]: the manufacturer. The vehicle keeper's liability insurer may have recourse to the manufacturer.[23]

Under German law, the manufacturer is subject to strict product liability from the Product Liability Act (ProdHaftG) and the fault-based tortious producer liability of Sect. 823 et seq. BGB. The Product Liability Act incorporates Directive 85/374/EEC into German Law (A new EU Directive on liability for defective products is going through the EU legislative process at the time of the writing of this article, see Procedure 2022/0302/COD.) It links liability for damages to the defectiveness of the product placed on the market. Sect. 3 of the Act legally defines a "defect" to exist if the product does not provide the safety that can reasonably be expected, taking into account all circumstances (para. 1). The assessment of what is a "reasonable expectation" should be based on "whether the product offers the safety which the generally accepted standards in the relevant area consider to be necessary".[24] The "generally accepted standards" are not to be determined empirically. Rather, it needs to be assessed "objectively"[25] what is to be regarded as "reasonable" from the point of view of users and third parties. On the one hand, this means that certainly no survey is to be held. On the other hand, it is still of some relevance what is already

---

[21] Term used in accordance with [85, p. 153 et seq.]. See also [64, pp. 465–66].

[22] Likewise at [26, p. 144].

[23] Cf. [2, 19, 44].

[24] BGH [= German Federal Court of Justice] NJW 2014, p. 2106 (2107 para. 8); NJW 2009, p. 2952 (2953 para. 12).

[25] Ibid. See also [76, pp. 97–98].

recognized as socially adequate among the relevant group [31, marginal no. 32 et seq.]. This includes hazards which, though familiar, are tolerated in the light of the product's benefits [31, marginal note 33]. The hazard from motor vehicles that remains despite extensive traffic regulation has so far been regarded as socially adequate.[26]

This view could change with respect to autonomous driving. Such is indicated by the road safety expectations communicated by responsible bodies. In 2020, 18,800 people lost their lives in road traffic in Europe,[27] and in 2021, 2569 in Germany.[28] The European Commission is now pursuing a "vision zero" for the year 2050 [20] (para. no. 45 as well as Annex I no. 16). The Union legislator recognises the potential of automated vehicles to "make a huge contribution to reducing road fatalities, as it is estimated that human error plays a role in more than 90% of road accidents".[29] The report of the Ethics Commission on Automated and Connected Driving, appointed by the German Federal Minister of Transport and Digital Infrastructure, states as its first ethical rule that "[t]he first purpose of partially and fully automated traffic systems [is] to improve the safety of all road users" [19, p. 10]. And the German road traffic lawmakers also intended to increase road safety with their 2017 amendment. In other words, driving systems are expected to drastically "do better" than humans.[30]

With respect to the future evolution of innovation law, an expectation of greater safety forges a double-edged sword. On the one hand, it can generate market demand for autonomous vehicles and political support. On the other hand, the generally accepted safety standard may become one that considers autonomous driving (largely) accident-free, which would shift the liability risk to the manufacturer and therefore have an inhibiting effect on innovation.[31] This is because driving errors—which could then be regarded as design flaws within the meaning of Sect. 3 ProdHaftG in case of such a generally accepted safety standard—will inevitably also occur with autonomous driving due to the autonomy risk. Such a development would be regrettable. If a certain "unpredictability" is inherent to an autonomous system (*supra* 2.2.2.2.1), not every driving error immediately implies a design error [54, p. 235], cf. [57, 64, 68]—a finding that is common knowledge for human drivers: Misjudgements and/or behaviour inappropriate to the traffic situation even happen to experienced and careful drivers without promptly resulting in the driver's licence being revoked due to a "design flaw" on the part of the driver (his or her unsuitability to drive a motor vehicle) [54, p. 235].

---

[26] See the references in footnote 40.

[27] https://ec.europa.eu/commission/presscorner/detail/en/IP_21_1767 (03/26/22).

[28] https://www.destatis.de/DE/Presse/Pressemitteilungen/2022/02/PD22_076_46241.html (03/26/22).

[29] Recital 23 of Regulation (EU) 2019/2144 of the European Parliament and of the Council of 27 Nov. 2019 on type-approval of motor vehicles […] (OJ L 325, 16 Dec. 2019, p. 1).

[30] Bundestag [German Federal Parliament] printed matter of 20 Feb. 2017, p. 13, referring to the Federal Government's Strategy for Automated and Connected Driving of 2015, there at 3.2.

[31] Dissenting: [26, pp. 143–44].

It is true that Sect. 1(2) no. 5 ProdHaftG excludes from product liability defects which could not have been known based on the state of science and technology at the time when the manufacturer placed the product on the market. However, the autonomy risk as such *is* known. Only the form of its manifestation in the individual case is unpredictable. It is therefore difficult to assess whether courts will assume an exclusion of liability under Sect. 1(2) no. 5 ProdHaftG. Legislative action therefore seems advisable as a precautionary measure to avoid an innovation obstacle.[32,33] The Conference of German Federal and State Ministers of Justice, however, merely produced the cryptic statement that even in the case of the use of autonomous systems, Sect. 1(2) no. 5 ProdHaftG still holds in the areas examined by the Conference.[34]

Finally, it could be considered identifying the autonomy risk (when materialising in individual cases) with a programming error. Programming errors are deemed to be *practically* unavoidable, and therefore a generally accepted standard is not presumed to exist that would require software to be completely free of programming errors. Programming errors therefore do not always prompt product liability.[35] However, such identification seems questionable. While programming errors could at least *theoretically* have been detected or avoided in advance, and can be remedied after their discovery, the autonomy risk is an inherently unavoidable risk of autonomous systems.[36]

The Union legislator could at least make it more difficult for a generally accepted standard of complete safety to emerge[37] by imposing realistic safety requirements on autonomous vehicles that do not explicitly call for a complete absence of driving errors. Of course, such requirements reasonably could still go far beyond a human driver's abilities.

2.2.2.3.2 Type-Approval of Autonomous Vehicles

A coinciding problem exists for the official type-approval of autonomous vehicles.

Fundamental rights oblige public authorities to protect people against harm from third parties—in this case, in the form of the placing on the market of harmful technologies.

---

[32] For this issue cf. [57, 67]. In detail [30, p. 581 et seq., 3, p. 46].

[33] Cf. [45, p. 348].

[34] 90th Conference of Federal Ministers of Justice, Report of the Länder Working Group on Digital Relaunch, No. 2 lit. c.

[35] [76, pp. 98–99], but commenting that, because of the serious hazards emanating from motor vehicles, software testing must be carried out with particular care.

[36] It is worth noting that the software of autonomous systems can suffer from traditional programming errors as well, cf. [72, S. 356]. Training a self-learning system with evidently inadequate data is likely to be considered such a traditional programming error, and, consequently, a design flaw, cf. [54, S. 235].

[37] Public law regulations on the approval of a technology do not determine civil liability. However, they can at least serve as an indication as to what seems to be the generally accepted standard, cf. BGH [= German Federal Court of Justice] NJW 2008, p. 3778 (3779 marginal no. 16); [25, marginal no. 25–26, 81, marginal no. 958 et seq.]. Cf. for autonomous driving: [41, 76].

Specifically, European harmonised vehicle type-approval law[38] needs to comply with Art. 2(1) (right to life) and Art. 3(1) (right to physical integrity) of the Charter of Fundamental Rights of the European Union (CFR).

In view of the high number of road deaths and serious injuries, a total ban of motor vehicles, being a particularly harmful technology, might initially appear to be called for. To highlight this point, imagine if the envisaged "air taxis" were expected to cause such a number of casualties. It is obvious that those would not be introduced in the first place.

However, for motor vehicle traffic the so-called "general risk of life" comes into play. This term encapsulates the idea that the existence of some familiar hazard in everyday life is deemed socially adequate based on an assumption that any additional action to further curb the hazard would unreasonably restrict free society. Motor vehicle traffic represents an everyday phenomenon that has been practised for over a 100 years; more than that, it shapes everyday life. A notable further increase in safety beyond existing regulations[39] could in principle only be achieved by drastic traffic restrictions, which would be incompatible with the idea of individual freedom. The remaining general risk of traffic accidents is therefore a general risk of life to be accepted (at least, this is the predominant view in the judicature and in the literature[40]; in light of the number of traffic fatalities and injuries, this view appears to be quite debatable). Therefore, motor vehicles are legally admissible.

At first glance, it would seem all the more appropriate to deem the remaining hazard of level 5 autonomous driving a general risk of life. After all, a further increase in road safety is expected.[41] However, it is doubtful if there really exists a willingness to accept a risk of accidents resulting from driving errors on the part of autonomous driving systems to the same extent as is the case for human error-based accidents.[42] This is because there is a lack of social familiarity with the risk of autonomous driving (in contrast to the traditional traffic accident risk including technical causes of accidents, such as tire blow-outs). As already discussed in relation to product liability (*supra* 2.2.2.2.1), it could therefore occur that social adequacy of the autonomy risk will not be assumed, or only to a very limited extent, and that as a result an extremely high type-approval hurdle is created.

To help avoid such an outcome, the legislator should state clearly that complete absence of driving errors cannot be achieved and is therefore not to be required from the manufacturer (see already *supra* 2.2.2.2.1). Within the limits of Union and national constitutional law, it is the responsibility of the democratic legislature to define the scope of socially adequate and thus permissible risks. The fact that already existing automated driving systems will not only be approved, but even be prescribed, is reason for optimism at present:

---

[38] In particular, Regulation (EU) 2019/2144 (fn. 29).

[39] Such as type-approval for motor vehicles, compulsory technical inspections, traffic rules.

[40] VG [= Administrative Court] Gelsenkirchen, Judg. of 23 May 2019–8 K 774/17, marginal no. 81; [58, marginal no. 75, 68, p. 551]. For criminal law: [36, pp. 68–69]. In detail: [18].

[41] Cf. [67, 68] and the references in the text in Sect. 2.2.2.3.1.

[42] Cf. [28, pp. 552–53, 569–70].

From 2022, Union law requires passenger cars and light commercial vehicles to be equipped with emergency braking and lane departure warning systems.[43]

2.2.2.3.3 Vehicle Autonomy Versus Criminal and Administrative Offences Law

Assuming, for the sake of the argument, that autonomous vehicles will be type-approved despite the autonomy risk, another challenge will arise: When an autonomous vehicle causes an accident with personal injury and/or property damage due to a traffic violation, there is no one criminally liable for this. The driver[44] as the user of the autonomous driving system is not liable for negligence.[45] The same applies to the manufacturer, at least insofar as "merely" the autonomy risk has materialized, which would have to be considered a permissible risk under the assumption made. Similarly, no party exists that could be held responsible for a traffic violation.[46]

The situation is therefore similar to purely natural causes of damage (e.g. death by lightning while outdoors). However, in the case of natural causes, there is no social expectation that someone should be prosecuted. In contrast, a motor vehicle is an artefact that is put into operation for the benefit of individual people. Doubts therefore appear justified as to whether society is prepared to accept that no one is held responsible in the case of serious traffic accidents [3, 26, 29], critic. [59, S. 401]. This forseeable criminal liability loophole could thus become an obstacle to innovation.[47] In contrast to civil liability and type-approval issues, the constitutional principle of culpability in criminal law makes it difficult to remedy the issue simply by legally assigning responsibility to someone, like the legislature has done, for example, in the case of strict liability of vehicle keepers.[48]

The situation is different for the manufacturer's obligation to continuously monitor its automated driving systems after their placing on the market.[49] Violating this obligation can constitute a criminal offense. This will be an additional motivation for the manufacturer to properly perform the monitoring and will thus make a remedy of identifable design flaws more likely (e.g., by providing software updates). Therefore, criminal liability, in this case, rather promotes than obstructs innovation.

---

[43] Articles 7 and 19 of Regulation (EU) 2019/2144 (fn. 29).

[44] At level 5, a human driver is not required at all.

[45] As long as he or she fulfills the specific duties that concern the use of the autonomous driving system.

[46] Cf. [36, pp. 68–69].

[47] On the other hand, it would also be an obstacle to innovation if manufacturers had to expect extensive criminal liability for driving errors [3, p. 47].

[48] [73, S. 397]. Thoughts on conceivable legislative responses at [13, p. 59 et seq.].

[49] On these in detail: [33].

2.2.3   Use of AI by Public Authorities

2.2.3.1 Current Situation and Prospects

Its superior ability to recognise patterns suggests a use of AI by public authorities as well. Mere decision support systems have long been in use, such as dialect recognition of asylum seekers.[50] Since mid-2019, the North Rhine-Westphalian judiciary has been undertaking a research project on an AI examination of seized files for child pornography content. Human evaluators reach their limits in the face of the amount of data to sift through, and technology thus speeds up law enforcement considerably.[51] Automatically pre-selected files likely presenting a criminal offense will, however, be subject to final human examination.

In contrast, conceivable future AI use could be designed to conclusively[52] determine whether the elements of a legal provision are met, or what the ensuing legal consequences are. Tremendous research efforts exist to "emulate" text comprehension and legal reasoning algorithmically [5, 24, 75, 81, 82]. Admittedly, there are also critical voices from legal scholarship on the prospects of success [6, 39, p. 15 et seq., 40, p. 140, 42, 43, 74, pp. 120–21, 77, p. 239].

At present, however, Sect. 35a of the German Administrative Procedure Act (VwVfG) prohibits[53] such extensive use of AI, at least on the legal consequences side. But it does admit an automated decision on whether the elements of a legal provision are met. However, the legislator had in mind only simple cases.[54] That rules out AI use for applying legal provisions whose elements demand an elaborate examination and weighing of the facts. And Art. 22 of the European General Data Protection Regulation (GDPR) and Art. 11 of Directive (EU) 2016/680[55] prohibit "automated individual decision-making" (exceptions are, however, admitted).

However, in view of the achievable improvements in the quality of decision-making, this conservative legal approach will be unsustainable. For example, AI decision-making

---

[50] Bundestag [German Federal Parliament] printed matter 19/190 of 8 Dec. 2017; [9, p. 14].

[51] https://www.land.nrw/de/pressemitteilung/kuenstliche-intelligenz-im-kampf-gegen-kinderpornographie; https://news.microsoft.com/de-de/ki-im-einsatz-gegen-kinderpornografie/ (03/26/22).

[52] Subject to review in individual cases by superiors or courts (as is also the case with decisions made by human officials).

[53] According to Sect. 155 para. 4 AO [Fiscal Code of Germany] and Sect. 31a SGB X [Code of Social Law, Book X] for tax law and for social law.

[54] Bundestag [German Federal Parliament] printed matter 18/8434 of 11 May 2016, p. 122; [71, p. 26].

[55] Regulation (EU) 2016/679 of the European Parliament and of the Council of 27 April 2016 on the protection of individuals with regard to the processing of personal data […] (OJ L 119, 4 May 2016, p. 1); Directive (EU) 2016/680 of the European Parliament and of the Council of 27 April 2016 on the protection of individuals with regard to the processing of personal data by competent authorities for the purposes of prevention, investigation, detection or prosecution of criminal offences or the execution of criminal penalties […] (OJ L 119, 4 May 2016, p. 89).

will avoid human bias, it will possess a (humanly impossible) complete overview of all legally relevant data, and it will provide an increased uniformity of the application of the law.[56]

2.2.3.2 Unpredictability as a Limit to AI Use?

However, the more AI itself "applies" the law, with the "man in the loop" disappearing, the more it must itself be able to fulfil the specific legal prerequisites for the exercise of government authority. These prerequisites include respect for the principles of democracy and the rule of law. The principle of democracy requires that official decisions be traceable to the people, just as the rule of law principle requires that those decisions be traceable to a legal basis. Both requirements are served by binding the exercise of government authority to democratically adopted law. The "unpredictability" of AI (*supra* para. 2.2.2.1) could jeopardize this bond. However, it is the conscientious human official who defines the standard of reliability to be applied. Similar to autonomous driving, it is to be expected that AI will at least match or probably even surpass humans. This is because human officials, too, are susceptible to making unlawful decisions based on errors or on extraneous motivation, which not even the most careful personnel selection and training could ever rule out. Restricting the use of AI for exercising government authority on the grounds of the technology's "unpredictability" could therefore only be based on a careful comparative analysis of human/AI error-proneness [52, p. 221 et seq., 54, pp. 237–38, 84, p. 44, 54].

2.2.3.3 Intransparency as a Limit to AI Use?

The principle of the rule of law imposes a further prerequisite on the exercise of government authority, which is particularly important when it comes to an interference with individual rights: decisions must be legally comprehensible and verifiable. To this end, public authorities must supplement its decisions with a statement of reasons and thus attempt to justify them (see, in particular, Sect. 39 VwVfG).

However, the algorithmic path that a self-learning AI takes to reach a result cannot be made transparent in detail. It results from what may be a very large number of successive layers of analysis that the system generates from the training data, eluding simple representation in the form "the system got from A via B to C".[57] For this reason, "algorithm control" is receiving considerable jurisprudential attention [4, 7, 12, 50, 84, p. 42 et seq.].

On the part of computer science, attempts are being made to meet this challenge with research into "Explainable AI" (XAI). This is not about extracting the actual problem-solving algorithm (this would not mean a legally sufficient "justification" anyway). Rather, it is to provide some sort of "secondary coding" of the algorithm to serve as an explanation. These efforts will still need time before they are ready for use. It is unclear, for

---

[56] On these hoped-for qualities: [27, 34, 59].
[57] Cf. [15, 17].

example, what is to be understood by an "explanation" of the algorithmic scheme or of the underlying models [17, 62, 70]. At least it is already possible to specify the data segment that was crucial for the system's response.

Assuming a future success of these efforts, such a secondary coding could be legally sufficient. Distinguishing between the production and the representation of a decision is already very familiar to jurisprudence[58] and does not imply an unlawful pathology from the outset. The actual mental and external processes that ultimately lead to an administrative decision are not to be equated with the offial statement of reasons. Rather, the statement of reasons condenses those processes into a conclusion as to the proper application of the law. Thus, there is some similarity between official reasons provided by human officials and XAI's "secondary coding" of the actual internal algorithm. Moreover, only in the case of human application of the law there may exist political considerations or a conscious or subconscious decision bias that do not enter into the official statement of reasons.

To conclude, AI may match or even surpass the human standard with respect to decision transparency as well, depending on further developments in information technology. Particularly in the case of deliberation processes whose outcomes are not fully determined by law (statutes that allow administrative discretion), AI may one day provide better reproducibility and—with XAI—a more precise explanation of the results.

From the current point of view, it would therefore appear inappropriate to consider the requirement of providing official reasons as a permanent obstacle to AI use by public authorities.

2.3   Regulatory Perspectives

2.3.1   Requirements on AI Regulation[59]

Present flaws in the judicial application of technology law suggest that future AI legislation should address more clearly the following two aspects.

2.3.1.1 Relevance of Third Party Benefits from AI for Fundamental Rights

Facially, technology regulation involves balancing economic fundamental rights of the manufacturers and the distributors of the technology (Art. 15 et seq. CFR, Art. 12 Basic Law for the Federal Republic of Germany [GG]) and fundamental rights of those whom the technology could cause harm, such as the rights to data protection (Art. 8 CFR, Art. 16 TFEU) and to physical integrity (Art. 2(1) and Art. 3(1) CFR, Art. 2(2) first sentence GG).

However, new technologies benefit not only manufacturers and distributors, but—in principle—third parties as well. This is certainly recognised by regulators and is

---

[58] [49, p. 51 et seq.] See also [37, margin no. 15, 30 et seq., 38, margin no. 47b et seq.].

[59] On this issue in further detail [51, p. 468, 53, p. 63 et seq., 54, p. 233 et seq.], [56] (Sect. 2).

sometimes used as a *political* argument for advocating new technologies (e.g., fewer road deaths thanks to autonomous driving). However, insufficient regard has been had to obstructive technology regulation *legally* infringing on fundamental rights of those who would otherwise enjoy third-party benefits (e.g., patients who die because a life-saving medical AI became available too late). Uncertainty as to whether the new technology would have really been available earlier in the absence of regulation does not lend itself as an argument against legal relevance of regulation-induced third-party harm. This follows from a rational application of the precautionary principle.[60] The principle's main purpose is to allow regulation even before a technology's potential for harm has been fully demonstrated.[61] Conversely, however, harm that could be caused *by regulation itself* must then, despite its uncertainty, also be relevant for legal consideration under fundamental rights.

Third-party harm *and* benefit must therefore be taken into account not only politically but also *legally*.

2.3.1.2 Use-Specific Risk Assessment and Regulation

Technology regulation requires a prior risk assessment. Internationally, a so-called "*risk-based* approach" is pursued, to which the EU subscribes, too.[62] The approach has two elements. First, there needs to be scientific indication of a technology's potential to cause harm *(hazard)*. Second, there needs to be some probability that the legal asset to be protected by the envisaged regulation could actually be exposed to this hazard in view of the planned use of the technology *(impact)*. Therefore, a use-specific assessment *(impact assessment)* will always be required.

However, following an unclear pattern, the Union shows an occasional tendency towards *hazard-based regulation*, of which the process- rather than product-oriented regulation of genetic engineering in the food sector provides the most evident example. And the German Federal Constitutional Court, too, allows the legislature to impute an unsubstantiated "basic risk" to genetic engineering that supposedly justifies regulation.[63] Technology regulation that displays such irrational neglect for the impact aspect violates the rule of law and fundamental rights of manufacturers, distributors, and, possibly, third parties (see 2.3.1.1). It is thus an obstacle to innovation.

As a conclusion, particular care must be taken to ensure that hazards from AI are only assessed and regulated on a *use-specific basis*, having regard to the individual impact of the technology. Unfortunately, with its Proposal for a Regulation "laying down

---

[60] General principle of Union law (case law), and in particular Art. 191(2) first subparagraph TFEU (environment).

[61] ECJ, judg. of 9 June 2016, Cases C-78/16 and C-79/16—Pesce and Serinelli v Presidenza del Consiglio dei Ministri and Others, para. 47, establishes practice of the courts.

[62] See, for example, EGC, judg. of 9 Sep. 2011, Case T-475/07—Dow AgroSciences and Others v Commission, [2011] ECR II-5937, para. 146 et seq.

[63] BVerfGE [German Federal Constitutional Court Reports] 128, 1 (39).

harmonized rules on artificial intelligence" (Artificial Intelligence Act) from April 2021,[64] the European Commission is presently pursuing a contrary approach: The Regulation would impose a plethora of standardized obligations on providers, importers, distributors and users of any AI system the European Union deems "high-risk".

2.3.2 Current Regulatory Efforts

The European Commission considers AI to be one of the strategically most important technologies of the twenty-first century and wants to ensure competitiveness of the Union [22, p. 2]. Within ten years, public and private investments across the Union are to increase to twenty billion euros per year [21, p. 3]. The German Federal Government is pursuing the "Artificial Intelligence Strategy" [10, p. 38 et seq.]. The European Union [European Commission Proposal, see Sect. 2.3.1.2 above], as well as German Federal and State Governments, are taking into consideration the existence of a complementary need for regulating AI's potential for harm [22, p. 17 et seq., 10, p. 38 et seq., 11 at 2.9] and they are using ethical guidelines from expert panels[65] for orientation. In the following, we take a closer look at the expert opinion of the Data Ethics Commission of the German Federal Government (DEK) established in 2018. The Commission devises ethical standards and guidelines as well as concrete recommendations for action on the topics of "data" and "algorithmic systems" [14, p. 34].

The largely balanced and nuanced nature of the Commission's findings and recommendations is to be appreciated.

The expert opinion correctly distinguishes the data perspective from the algorithm perspective, whose discourses complement and necessitate each other—which is exactly why they should analytically be kept apart (p. 36, see also pp. 26–27, 184). Although AI is generally characterized by a large amount of processed data, this also applies to other scenarios of electronic data processing. Consequently, violations of data protection rules do not present a potential for harm being specific to AI only [55, pp. 93–94].

The DEK advocates a risk-adaptive regulatory regime that aims to achieve a balance between the amount of potential for harm and the amount of burden from regulation. Severity and probability of harm are considered to partially depend on *actual impacts* from algorithm-based decisions (risk-based approach, supra 2.3.1.2) and on a state or private sector context. The role of the AI system as part of the decision-making process is also relevant: From a mere inspiration of human actors via a *de facto* commitment to automated recocommendations to a complete absence of a "man in the loop" [14, p. 161,

---

[64] European Commission, Proposal for a Regulation of the European Parliament and of the Council Laying Down Harmonized Rules on Artificial Intelligence" (Artificial Intelligence Act), 21 April 2021, COM(2021) 206 final.

[65] For the European Commission: Independent High Level Expert Group on Artificial Intelligence, Ethics Guidelines for Trustworthy AI, 2019.

174]. A "criticality pyramid" is introduced to signify potential for harm from algorithmic systems along levels 1 to 5. Level 1 does not require regulation, while level 5 AI use should be banned (pp. 173–174, 177).

The individual regulatory response should be able to pick instruments from a rich toolbox, accounting for the varying contexts of AI use and vulnerabilities (pp. 166, 175). With regard to transparency requirements, the research on "Explainable AI" is welcomed (pp. 169–170, 175, 187) (see *supra* 2.2.3.3).

The DEK acknowledges that human decision-making, too, is subject to error and bias, and it considers this fact an opportunity for implementing superior technology-based processes (pp. 167–170). It even considers the use of an algorithmic system to be ethically imperative should the system achieve better results than humans in all ethically relevant aspects (p. 172). Consequently, it emphasizes that regulation should not obstruct technological and social innovations and dynamic market development (pp. 41, 173, 178). For use in public administration, safety requirements are deemed to generally be higher (at least criticality level 3), and the best available technology is called for (p. 212). On this basis, a cautious extension of the scope of application of Sect. 35a VwVfG (on this provision *supra* 2.2.3.1) is considered acceptable (p. 214). However, for areas where public administration has the authority to interfere with individual rights (esp. law enforcement), the DEK discourages AI use due to an alleged lack of controllability (p. 214). This stance seems too restrictive; the autonomy risk does not lend itself as a general objection, considering that human officials, too, are (likely even more) error-prone (see *supra* 2.2.3.2). Therefore, areas of public administration that involve an interference with individual rights should rather be a first choice for AI use as an effective tool for increasing decision-making quality.

Even beyond that, the report is to be criticised for its treatment of the autonomy risk, which is not systematically addressed. The DEK only states in general terms that algorithmic systems must function correctly and reliably (p. 25), and that malfunctions could be distinguished from normal operation in the majority of cases (p. 220). Thus, there is no useful guidance on how to deal with the inevitability of individual errors of self-learning systems.

The DEK acknowledges that "regulation of a general nature bears the risk that resulting obligations will be enforced even in cases where there is no sufficient potential for harm, for a horizontal legal act cannot sufficiently reflect the minute differences between risky and less risky applications existing in reality" (p. 181, author's translation). This is perfectly in line with the above conclusion that hazards from AI should only be assessed and regulated on a use-specific basis (*supra* 2.3.1.2). And yet, at the same time, the DEK does propose an EU Regulation on Algorithmic Systems (EUVAS), which would apply to private and public actors and would impose *general regulation* on the admissibility and the design of algorithmic systems with respect to system criticality, on transparency, on affected parties' rights and on a supervisory structure (p. 180 et seq.). With this proposal, the DEK departs to some extent from its nuanced approach that acknowledges the need for use-specific regulation (even though it emphasises that the EUVAS should be supplemented by detailed sectoral regulations).

For comparison, the regulatory challenge is different with the General Data Protection Regulation (GDPR). The GDPR exclusively protects the fundamental right to data protection.[66] In contrast, potential harm from AI is not limited to affecting only one specific fundamental right. Therefore, there exists no basis for prefacing use-specific regulation with a general regulatory part.

Overall, the DEK's findings evidently reflect a compromise between the different positions represented in the Commission. It deserves to be noted that technology-friendly positions, too, managed to be clearly included in the guidelines. In any case, policymakers could not convincingly rely on the DEK's expert opinion for an (overly) restrictive regulation of AI. Unfortunately, the proposed EU Artificial Intelligence Act (AIA, expected to enter into force still in 2024) falls short of matching the DEK's largely nuanced approach. An extremely wide array of areas of AI application is labeled 'high risk', prompting uniform applicability of severely restrictive requirements without any regard to the nature of the individual use sector and the hazards specific to it. Furthermore, one of those requirements is the general need for 'human oversight' (Article 14). This demonstrates that the European legislator neither sufficiently considered the potential superiority in quality of AI decision-making over humans, nor human error-proneness that could be circumvented by AI use. It is to be hoped that at least the harmonised standards and common specifications, to be adopted by European standardisation organisations and by the European Commission (Art. 40–1 AIA), will be more sector-specific.

References

1. Ahlberg H (2019) § 2 UrhG (Stand: 20.04.2018). In: ders., Götting H-P (Hrsg) BeckOK Urheberrecht, 26. Edition (15.10.2019), München
2. Armbrüster C (2017) Automatisiertes Fahren—Paradigmenwechsel im Straßenverkehrsrecht? Z Rechtspol (ZRP), 83–86
3. Beck S (2020) Die Diffusion strafrechtlicher Verantwortlichkeit durch Digitalisierung und Lernende Systeme. Z Internation Strafrechts (ZIS), 41–50
4. Berger A (2017) Digitales Vertrauen—eine verfassungs- und verwaltungsrechtliche Perspektive. Deutsches Verwaltungsblatt (DVBl), 804–808
5. Branting K (2017) Data-centric and logic-based models for automated legal problem solving. Artif Intell Law 25:5–27
6. Buchholtz G (2017) Legal Tech. Chancen und Risiken der digitalen Rechtsanwendung. Juristische Schulung (JuS), 955–960
7. Bull HP (2017) Der "vollständig automatisiert erlassene" Verwaltungsakt—zur Begriffsbildung und rechtlichen Einhegung von "E-Government". Deutsches Verwaltungsblatt (DVBl), 409–417
8. Bullinger W (2019) § 2 UrhG. In: ders., Wandtke A-A (Hrsg) Praxiskommentar Urheberrecht, 5. Aufl. Beck, München

---

[66]Art. 8(1) CFR and Art. 16(1) Treaty on the Functioning of the European Union (TFEU).

9. Bundesamt für Migration und Flüchtlinge (BAMF) (Hrsg) (2017) Digitalisierungsagenda 2020, 4. Fassung 4/2019, Nürnberg, https://www.bamf.de/SharedDocs/Anlagen/DE/Digitalisierung/broschuere-digitalisierungsagenda-2020.pdf?__blob=publicationFile&v=11. Accessed 26 Mar 2022

10. Bundesregierung (2018) Strategie Künstliche Intelligenz, https://www.bmwi.de/Redaktion/DE/Publikationen/Technologie/strategie-kuenstliche-intelligenz-der-bundesregierung.pdf?__blob=publicationFile&v=8. Accessed 26 Mar 2022

11. Bundesregierung (2019) Zwischenbericht ein Jahr KI-Strategie. https://www.bmas.de/SharedDocs/Downloads/DE/Pressemitteilungen/2019/ki-ein-jahr-zwischenbericht.pdf;jsessionid=17993E02A426A2D775DE7F05860D785C.delivery1-master?__blob=publicationFile&v=1. Accessed 26 Mar 2022

12. Bundestag, Wissenschaftliche Dienste (2017) Algorithmen. Einzelfragen zu Instrumenten und Regelansätzen, WD 8–3000—031/17. Berlin

13. Cornelius K (2020) "Künstliche Intelligenz", Compliance und sanktionsrechtliche Verantwortlichkeit. Z Internation Strafrechts (ZIS), pp 59–64

14. Datenethikkommission der Bundesregierung (DEK) (2019) Gutachten. Berlin

15. Desai DR, Kroll JA (2017) Trust but Verify. Harv J Law Tech 31:1–64

16. Dornis TW (2019) Der Schutz künstlicher Kreativität im Immaterialgüterrecht. Gewerblicher Rechtsschutz und Urheberrecht (GRUR), 1254–1264

17. Edwards L, Veale M (2017/2018) Slave to the algorithm: why a right to an explanation is probably not the remedy you are looking for. Duke L Tech Rev 16:18–84

18. Martin E (1995) Ist der Straßenverkehr in seiner heutigen Form verfassungswidrig? "Verstößt das bestehende Recht des Straßenverkehrs gegen den Lebens und Gesundheitsschutz des Grundgesetzes?". Verwaltungsblätter für Baden-Württemberg (VBlBW), 161–172

19. Ethik-Kommission Automatisiertes und Vernetztes Fahren (2017) Bericht. https://www.bmvi.de/SharedDocs/DE/Publikationen/DG/bericht-der-ethik-kommission.pdf?__blob=publicationFile. Accessed 26 Mar 2022

20. Europäische Kommission (2011) Weißbuch. Fahrplan zu einem einheitlichen europäischen Verkehrsraum—Hin zu einem wettbewerbsorientierten und ressourcenschonenden Verkehrssystem. KOM(2011) 144 endgültig vom 28.3.2011

21. Europäische Kommission (2018) Koordinierter Plan für künstliche Intelligenz. COM(2018) 795 final vom 07.12.2018

22. Europäische Kommission (2018) Künstliche Intelligenz für Europa. COM(2018) 237 final vom 25.04.2018

23. Europäisches Parlament (2017) Bericht des Rechtsausschusses mit Empfehlungen an die Kommission zu zivilrechtlichen Regelungen im Bereich Robotik. 2015/2103(INL) vom 27.01.2017. Plenarsitzungsdokument A8–0005/2017

24. Feteris ET (Hrsg) (2017) Fundamentals of legal argumentation. a survey of theories on the justification of judicial decisions, 2. Aufl. Springer, Dordrecht

25. Förster C (2020) § 3 ProdHaftG. In: Bamberger HG, Roth H, Hau W, Poseck R (Hrsg) Beck'scher Online-Kommentar (BeckOK) BGB. Stand 1.2.2020. München

26. Fraunhofer-Institut für Arbeitswirtschaft und Organisation IAO (2014) Hochautomatisiertes Fahren auf Autobahnen—Industriepolitische Schlussfolgerungen. Studie im Auftrag des Bundesministeriums für Wirtschaft und Energie

27. Fries M (2016) PayPal Law und Legal Tech—Was macht die Digitalisierung mit dem Privatrecht? Neue Juristische Wochenschrift (NJW), 2860–2865

28. Gasser TM (2015) Grundlegende und spezielle Rechtsfragen für autonome Fahrzeuge. In: Maurer M, Gerdes JC, Lenz B, Winner H (eds) Autonomes Fahren. Technische, rechtliche und gesellschaftliche Aspekte. Springer, Berlin/Heidelberg, pp 543–574

29. Gless S, Janal R (2016) Hochautomatisiertes und autonomes Autofahren—Risiko und rechtliche Verantwortung. Juristische Rundschau (JR), 561–575
30. Gleß S, Weigend T (2014) Intelligente Agenten und das Strafrecht. Z Gesamte Strafrechtswissenschaft (ZStW) 126:561–591
31. Goehl Benjamin (2019) § 3. In: Spickhoff A (Hrsg) beck-online.GROSSKOMMENTAR (BeckOGK). ProdHaftG, München, Stand 1.9.2019
32. Gomille C (2019) Kreative künstliche Intelligenz und das Urheberrecht. JuristenZeitung (JZ) 969–175
33. Gortan M (2018) Unterlassensstrafbarkeit geschäftsleitender Personen des Softwareherstellers selbstfahrender Fahrzeuge durch Produktbeobachtungspflichtverletzung. Computer und Recht (CR), 546–552
34. Guggenberger L (2019) Einsatz künstlicher Intelligenz in der Verwaltung. Neue Z für Verwaltungsrecht (NVwZ), 844–850
35. Hetmank S, Lauber-Rönsberg A (2018) Künstliche Intelligenz—Herausforderungen für das Immaterialgüterrecht. Gewerblicher Rechtsschutz und Urheberrecht (GRUR), 574–582
36. Hilgendorf E (2018) Automatisiertes Fahren und Strafrecht—der "Aschaffenburger Fall". Deutsche Richterzeitung (DRiZ), 66–69
37. Hoffmann-Riem W (2012) Eigenständigkeit der Verwaltung. In: ders., Schmidt-Aßmann E, Voßkuhle A (Hrsg) Grundlagen des Verwaltungsrechts (GVwR), Bd. I. 2. Aufl., München, § 10
38. Hoffmann-Riem W (2012) Rechtsformen, Handlungsformen, Bewirkungsformen. In: Hoffmann-Riem W, Schmidt-Aßmann E, Voßkuhle A (Hrsg) Grundlagen des Verwaltungsrechts (GVwR), Bd. I. 2. Aufl., München, § 33
39. Hoffmann-Riem W (2017) Verhaltenssteuerung durch Algorithmen—eine Herausforderung für das Recht. Archiv des öffentlichen Rechts (AöR) 142:1–42
40. Hoffmann-Riem W (2019) Die digitale Transformation als Herausforderung für die Legitimation rechtlicher Entscheidungen. In: Unger S, von Ungern-Sternberg A (eds) Demokratie und künstliche Intelligenz. Mohr Siebeck, Tübingen, pp 129–159
41. Jänich VM, Schrader PT, Reck V (2015) Rechtsprobleme des autonomen Fahrens. Neue Z für Verkehrsrecht (NZV), 313–318
42. Kotsoglou KN (2014) Schlusswort. "Subsumtionsautomat 2.0" reloaded?—zur Unmöglichkeit der Rechtsprüfung durch Laien. JuristenZeitung (JZ), 1100–1103
43. Kotsoglou K N (2014) Subsumtionsautomat 2.0. Über die (Un-)Möglichkeit einer Algorithmisierung der Rechtserzeugung. JuristenZeitung (JZ), 451–457
44. Kütük-Markendorf ME, Essers D (2016) Zivilrechtliche Haftung des Herstellers beim autonomen Fahren. Haftungsfragen bei einem durch ein autonomes System verursachten Verkehrsunfall. Zeitschrift für IT-Recht und Recht der Digitalisierung (Multimedia und Recht—MMR), 22–26
45. Lange U (2017) Automatisiertes und autonomes Fahren—eine verkehrs-, wirtschafts- und rechtspolitische Einordnung. Neue Z für Verkehrsrecht (NZV), 345–352
46. Lauber-Rönsberg A (2019) Autonome "Schöpfung"—Urheberschaft und Schutzfähigkeit. Gewerblicher Rechtsschutz und Urheberrecht (GRUR), 244–253
47. Legner S (2019) Erzeugnisse Künstlicher Intelligenz im Urheberrecht. Z für Urheber- und Medienrecht (ZUM), 807–812
48. Lüdemann V, Sutter C, Vogelpohl K (2018) Neue Pflichten für Fahrzeugführer beim automatisierten Fahren—eine Analyse aus rechtlicher und verkehrspsychologischer Sicht. Neue Z für Verkehrsrecht (NZV), 411–417
49. Luhmann N (1966) Recht und Automation in der öffentlichen Verwaltung. Duncker & Humblot, Berlin
50. Martini M, Nink D (2017) Wenn Maschinen entscheiden …. Neue Z für Verwaltungsrecht—extra (NVwZ—extra) 10/2017, S. 1–14

51. Meyer S (2011) Risikovorsorge als Eingriff in das Recht auf körperliche Unversehrtheit. Gesetzliche Erschwerung medizinischer Forschung aus Sicht des Patienten als Grundrechtsträger. Archiv des öffentlichen Rechts (AöR) 136:428–478
52. Meyer S (2014) Der Einsatz von Robotern zur Gefahrenabwehr. In: Hilgendorf E (Hrsg) Robotik im Kontext von Recht und Moral. Reihe "Robotik und Recht", Bd 3. Nomos, Baden-Baden, S. 211–237
53. Meyer S (2015) "Gefühlsschutz" und hazardbasierte Vorsorge im Unionsrecht als Herausforderung des Rechtsstaatsprinzips. In: Möstl M (Hrsg) Lebensmittelanalytik und Recht. Schriftenreihe zum Lebensmittelrecht, Bd 35. Bayreuth, S. 63–107
54. Meyer S (2018) Künstliche Intelligenz und die Rolle des Rechts für Innovation. Rechtliche Rationalitätsanforderungen an zukünftige Regulierung. Z für Rechtspolitik (ZRP), 233–238
55. Meyer S (2019) Sekundärrechtliche Vorgabe einer Verwendung Künstlicher Intelligenz bei der mitgliedstaatlichen Durchführung des Unionsrechts. Ihre Vereinbarkeit mit der sogenannten "Verfahrensautonomie" und mit materiellrechtlichen Anforderungen. Österreichische Z für Wirtschaftsrecht (ÖZW), 83–97
56. Meyer S (2020) "Theorie" und Praxis des Risikorechts. In: Opper J, Rolfes V, Roth PH (eds) Chancen und Risiken der Stammzellforschung. Wissenschafts-Verlag, Berlin, pp 279–292
57. Müller-Hengstenberg CD, Kirn S (2014) Intelligente (Software-)Agenten: Eine neue Herausforderung unseres Rechtssystems. Rechtliche Konsequenzen der "Verselbstständigung" technischer Systeme. Zeitschrift für IT-Recht und Recht der Digitalisierung (Multimedia und Recht—MMR), 307–313
58. Müller-Terpitz R (2009) Recht auf Leben und körperliche Unversehrtheit. In: Isensee J, Kirchhof P (Hrsg) Handbuch des Staatsrechts. Bd VII. C.F. Müller, 3. Aufl., Heidelberg, § 147
59. Nehm K (2018) Autonomes Fahren—Bremsen Ethik und Recht den Fortschritt aus? JuristenZeitung (JZ), 398–402
60. Nordemann A (2018) § 2 UrhG. In: Nordemann A, Fromm JB, Czychowski C (eds) Urheberrecht, 12th edn. Beck, München
61. Ory S, Sorge C (2019) Schöpfung durch Künstliche Intelligenz? Neue Juristische Wochenschrift (NJW), 710–713
62. Páez A (2019) The pragmatic turn in explainable artificial intelligence (XAI). Mind Mach 29:441–459
63. Schirmer J E (2017) Augen auf beim automatisierten Fahren! Die StVG-Novelle ist ein Montagsstück. Neue Z für Verkehrsrecht (NZV), 253–257
64. Schirmer JE (2018) Robotik und Verkehr. Was bleibt von der Halterhaftung? Rechtswissenschaft (RW) 9:453–476
65. Schirmer JE (2019) Von Mäusen, Menschen und Maschinen—autonome Systeme in der Architektur der Rechtsfähigkeit. JuristenZeitung (JZ), 711–718
66. Schmid A, Wessels F (2017) Event Data Recording für das hoch- und vollautomatisierte Kfz—eine kritische Betrachtung der neuen Regelungen im StVG. Neue Z für Verkehrsrecht (NZV), 357–364
67. Schrader P T (2016) Haftungsfragen für Schäden beim Einsatz automatisierter Fahrzeuge im Straßenverkehr. Deutsches Autorecht (DAR), 242–246
68. Schulz T (2017) Sicherheit im Straßenverkehr und autonomes Fahren. Neue Z für Verkehrsrecht (NZV), 548–553
69. Schulze G (2018) § 2. In: Schulze G, Dreier T. Urheberrechtsgesetz, 6. Aufl. Beck, München
70. Sheh R, Monteath I (2018) Defining explainable AI for requirements analysis. KI—Künstliche Intelligenz 32:261–266
71. Siegel T (2017) Automatisierung des Verwaltungsverfahrens—zugleich eine Anmerkung zu §§ 35a, 24 I 3, 41 IIa VwVfG. Deutsches Vewaltungsblatt (DVBl), 24–28

72. Singler P (2017) Die Kfz-Versicherung autonomer Fahrzeuge. Neue Z für Verkehrsrecht (NZV):353–357
73. Staub C (2019) Strafrechtliche Fragen zum Automatisierten Fahren. Der Hersteller als strafrechtlicher Verantwortlicher der Zukunft?—Umfang der Sorgfaltspflicht—Datenschutz versus Aufklärungspflicht. Neue Z für Verkehrsrecht (NZV), 392–398
74. Unger S (2019) Demokratische Herrschaft und künstliche Intelligenz. In: Unger S, von Ungern-Sternberg A (eds) Demokratie und künstliche Intelligenz. Mohr Siebeck, Tübingen, pp 113–128
75. Verheij B (2017) Proof with and without probabilities. Correct evidential reasoning with presumptive arguments, coherent hypotheses and degrees of uncertainty. Artif Intell Law 25:127–154
76. von Bodungen B, Hoffmann M (2018) Hoch- und vollautomatisiertes Fahren ante portas—Auswirkungen des 8. StVG-Änderungsgesetzes auf die Herstellerhaftung. Neue Z für Verkehrsrecht (NZV), 97–102
77. von Graevenitz A (2018) "Zwei mal Zwei ist Grün"—Mensch und KI im Vergleich. Z für Rechtspolitik (ZRP), 238–241
78. Wachenfeld W, Winner H (2015) Lernen autonome Fahrzeuge? In: Maurer M, Gerdes JC, Lenz B, Winner H (eds) Autonomes Fahren. Technische, rechtliche und gesellschaftliche Aspekte. Springer, Berlin/Heidelberg, pp 465–488
79. Wagner G (2017) § 823 BGB. In: Säcker FJ, Rixecker R, Oetker H, Limperg B (Hrsg) Münchener Kommentar zum Bürgerlichen Gesetzbuch. Bd 6, 7. Aufl. Beck, München
80. Walter A (2019) § 7. In: Spickhoff A (Hrsg) beck-online.GROSSKOMMENTAR (BeckOGK). StVG. Stand 1.9.2019. München
81. Waltl B et al (2019) Semantic types of legal norms in German laws. Artif Intell Law 27:43–71
82. Walton D, Sartor G, Macagno F (2016) An argumentation framework for contested cases of statutory interpretation. Artif Intell Law 24:51–91
83. Wiebe A (2019) § 2 UrhG. In: Spindler G, Schuster F (eds) Recht der elektronischen Medien, 4th edn. Beck, München
84. Wischmeyer T (2018) Regulierung intelligenter Systeme. Archiv des öffentlichen Rechts (AöR) 143:1–66
85. Zech H (2016) Zivilrechtliche Haftung für den Einsatz von Robotern—Zuweisung von Automatisierungs- und Autonomierisiken. In: Gless G, Seelmann K (Hrsg) Intelligente Agenten und das Recht. Nomos, Baden-Baden, S 153–204

Professor Stephan Meyer studied Public Law and Economics at the University of the German Federal Armed Forces Munich. He received his doctorate in 2003 with a thesis on the legislative powers of the Upper Chamber of Parliament. After being an assistant professor 2004–2010, he completed his habilitation at the University of Erfurt in 2010 and was awarded the Venia Legendi for Public Law, Administrative Science and Legal Theory. In 2016, he was appointed a Professor of Public Law at the Technical University of Applied Sciences Wildau. His research focuses on the law of innovation, risk regulation, and data protection law.

Part III

Application Potential

Application Potential of Artificial Intelligence in the Car Trade

3

Michael Gröschel, Gabriele Roth-Dietrich, and Carl-Christian Neundorf

Abstract

Social trends such as urbanization, digital transformation and the sharing economy are influencing the future of the automotive trade. Four mobility scenarios, in which cars are privately owned or shared on the one hand and are driven by a driver or move autonomously on the other, are changing the most important business processes in the automotive trade, such as car sales, maintenance, repairs and used car purchases. In all processes, application scenarios for the artificial intelligence (AI) subfields computer vision, natural language processing (NLP) and machine learning can be identified. The analysis shows that the current degree of implementation varies widely, from the already widespread predictive maintenance in maintenance, to the first approaches of computer vision for diagnosis in repair and used car purchase, to the still novel use of chatbots in consulting and machine learning in recommendation processes. For car dealers, the question arises of specialization, e. g. as a service factory or for the fleet management of a mobility service provider, each of which requires different AI scenarios.

M. Gröschel (✉) · G. Roth-Dietrich
Hochschule Mannheim, Mannheim, Germany
e-mail: m.groeschel@hs-mannheim.de; g.roth-dietrich@hs-mannheim.de; https://www.taxxas.com; https://www.informatik.hs-mannheim.de/wir/menschen/professoren/prof-dr-gabriele-roth-dietrich.html

C.-C. Neundorf
solute GmbH, Karlsruhe, Germany

© The Author(s), under exclusive license to Springer Fachmedien Wiesbaden GmbH, part of Springer Nature 2024
T. Barton, C. Müller (eds.), *Artificial intelligence in application*,
https://doi.org/10.1007/978-3-658-43843-2_3

3.1   Introduction

The German *automotive industry* is one of the main pillars of the German economy (see [1, p. 14]). In 2017, for example, it was responsible for around 20% of total industrial sales, with sales of around EUR 423 billion (cf. [2, p. 2]). This aspect is also reflected in the fact that the German *automotive market*, in terms of car sales, is one of the top markets in Europe (cf. [2, p. 3]). However, trends such as *urbanization, digitalization* as well as the *sharing economy* lead to the fact that more and more potential customers question the ownership of a car (cf. [3, p. 18]). This leads to the question of how the *automotive trade* should deal with these trends and how *artificial intelligence (AI)* can support the trade in adapting to the new circumstances. AI, although still very much in development as part of the digital transformation, is now omnipresent in everyday commercial life (cf. [4, p. 4]), whether through the use of the smartphone personal assistant, fraud detection or translation software such as DeepL, for example. In terms of the global economy, the use of AI could positively influence global economic growth by up to 20% by 2030 (cf. [4, p. 3]). On the one hand, this can be attributed to the increase in productivity that goes hand in hand with the use of AI (cf. [4, p. 8]). On the other hand, AI also influences the consumer side. This happens through improved product personalization and increased product quality (cf. [5, p. 11 f.]). This gives an idea of the potential that this technology holds from both an economic and a technical perspective.

This article examines which roles and tasks the automotive trade will assume as the digital transformation progresses and how AI can be used successfully and effectively in the automotive industry. To this end, Sect. 3.2 identifies and classifies challenges for the automotive trade, decisive social trends, but also possible scenarios for the future roles and tasks of the automotive trade on the basis of a literature analysis. In Sect. 3.3, the *core processes in the car trade,* which were identified among other things through expert interviews with the company aubex GmbH, are elaborated. These serve as a basis for developing possible use cases for AI. Section 3.4 contains a brief introduction to the relevant subfields of AI and highlights the economic effects associated with the use of AI. Building on these insights, Sect. 3.5 considers possible use cases of AI in car retailing based on grouping by service creation processes. Section 3.6 draws a final conclusion.

3.2   Challenges for the Automotive Trade

3.2.1   Social Trends

Social trends undoubtedly have an impact on the buying and selling behaviour of the population with regard to cars. They tend to override conventional developments and can set new directions (cf. [3, p. 13]). These trends include *urbanization, digitalization* and the resulting *sharing economy,* which is concretized in the automotive sector as *car sharing.*

3.2.1.1 Urbanisation

It can be assumed that by 2030 about 60% of the world's population will live in large cities (cf. [6, p. 96]). By 2050, this figure is expected to rise to between 70% and 80% of the global population, based on a figure of around 10–12 billion people (cf. [7, p. 11]). It can be deduced from this that life in very densely populated urban areas will become a typical form of existence for the majority of humanity in the twenty-first century. Thus, it is also foreseeable that due to the large number of people who have to arrange their diverse needs in an ever smaller area, the functional space for automobility as we know it will become increasingly scarce (cf. [7, p. 11]). This implies that there will be no more room in large cities for the ever-growing automobile fleets due to urbanization. Furthermore, urbaniza-tion causes a rethinking regarding the attitude towards cars. For example, cars are increas-ingly perceived as a nuisance in big cities because, in contrast to rural areas, alternative methods of transportation are available (cf. [3, p. 14]). For example, according to Parment ([3, p. 14]), parking and inspections as well as insurance are usually more expensive, road traffic is limited by various restrictions, and public transport is usually more flexible and faster. These circumstances lead to the fact that driving and thus also the ownership of a car are questioned by many.

As a result of this trend, the automotive industry is increasingly having to deal with the major effects of these social changes, so that a rethink in the direction of new technologies and mobility concepts appears necessary. A decisive role is played here by the increased environmental impact, the growing awareness of this, as well as the loss of time due to a slow traffic flow and traffic jams on the main traffic routes (cf. [6, p. 96]). In addition, the sharing economy, which is becoming increasingly popular among Generation Y, together with the constantly growing environmental awareness, is leading the urban population to rely more on mobility services instead of owning a car (cf. [6, p. 96]).

3.2.1.2 Digitisation and Digitalisation

The terms *digitisation* and *digitalisation* are not clearly defined and can be interpreted in many different ways (see [8]). If the term is interpreted from a technical perspective, it describes the conversion of analogue information into a digital form of storage (digitisa-tion). Traditionally, the term digitalisation is used to describe the transfer of tasks that were previously performed by humans to computers. Nowadays, the term digitalisation is often used as a synonym for the introduction of digital technologies in companies and as a driver of digital transformation (see [8]). The term *digital transformation* describes both the significant changes in everyday life, the economy and society through the use of digital technologies and techniques, as well as the effects that go hand in hand with this (cf. [9]).

According to Parment [3], the effect of digitalisation on the automotive industry is, on the one hand, that digitalisation has simplified the use of *intermodal mobility offers* and that there is thus a greater range of options for everyone on how they can get to a desired destination as cheaply and punctually as possible. On the other hand, this simplified use of intermodal mobility in combination with the trend of urbanization (and the effects of this trend) means that the car is becoming increasingly less important, especially among younger people, and this is leading to a shift in preferences from individual to collective modes of transport (cf. [3, p. 18]).

However, digitalisation and the use of big data can also make customers and their behavior more transparent for manufacturers and dealers, so that offers can be better individualized and better *customer segmentation can be* achieved at an aggregated level. Furthermore, the in-depth customer knowledge created by Big Data analyses can be used to adapt the range of services offered by both the new car business and mobility providers to the needs of the customer in line with the situation.

Hamidian and Kraijo see great potential in the *networking of* cars with the Internet and the workshop, which only became possible through digitalization (cf. [10, p. 10]). For example, telemetry could be used to transmit data from the car to the owner's preferred workshop and to carry out remote diagnoses. Furthermore, not only cars could be networked, but also car dealers and workshops among themselves, so that a kind of *"service and parts value network"* would be created [10]. Furthermore, if, for example, a remote diagnosis were to reveal that a part urgently needed to be replaced due to wear and tear but was not in stock at the workshop, this "service and parts value creation network" could be used to order the part from another workshop so that it could be quickly replaced at the customer's preferred workshop.

3.2.1.3 Sharing Economy

The term *sharing economy* refers to the systematic lending or renting of objects and the mutual provision of spaces and areas, primarily by private individuals and interest groups (cf. [11, 12]). Here, the focus of the sharing economy is not on the acquisition of ownership, but merely on the *temporary use of* other people's property (cf. [13, p. 245]). The rise and thus also the increasing importance of the sharing economy is related, among other things, to digitalization and social change, combined with the ever-increasing environmental awareness (cf. [14, p. 360 f.]). Social change in this respect is particularly evident among younger generations. Here, the change is characterized by the development of the consumer towards the so-called transumer, who prefers customer experience or time-limited access to goods to permanent possession. The consumer rejects possessions and property and increasingly strives for active experiences (cf. [14, p. 361]). Another factor that favours the spread of the sharing economy are the economic advantages that come with it (cf. [14, p. 362]). Thus, only the active use of the goods or services is paid for, without having to pay for additional fixed costs (cf. [11, p. 232]).

The aforementioned reasons that favour the growth of the sharing economy may also be a reason why *car sharing* is slowly but steadily gaining importance in Germany. According to the Bundesverband CarSharing (German CarSharing Association), the number of registered users has grown steadily from 116,000 in the beginning (2008) to 2.4 million (2019) users (cf. [15]). This is in contrast to a survey by IfD Allensbach (cf. [16]), according to which the number of users is only 1.08 million in 2019. However, this survey also confirms that interest in carsharing services exists and has grown. Thus, the interest in carsharing grew from 7.76 million (2015) to 9.29 million (2019) interested parties (see [16]). This growth is also reflected in the revenue forecasts. It is expected that carsharing will generate a turnover of 3.7–5.6 billion euros in 2020 or a global turnover of up to 9.1 billion US dollars in 2025 (cf. [17, 18]). Based on these forecasts, it can be seen that car sharing is a global trend with future potential.

Shaheen and Cohen provide a precise analysis of the consequences of this trend for the automobile market and, in particular, for the automobile trade (cf. [19]). Thus, car sharing together with increasing environmental awareness and increased use of alternative means of transport leads to a reduction in Vehicle Kilometres Travelled (VKT), consequently to a reduction in road traffic as well as to a reduction in vehicle ownership and to car sharing participants foregoing vehicle purchases (cf. [19, p. 19]). A carsharing car replaces between 4 and 10 private vehicles in Europe, between 9 and 13 in North America and between 7 and 10 in Australia (cf. [19, p. 9]). With regard to the German market, a survey by Trendmonitor Germany provides further insights (cf. [20]). Thus, 25% of the total population could imagine using car sharing instead of buying a car. In the group of 18–29 year-olds, 50% of respondents could imagine using car sharing and at the same time foregoing the purchase of a car. Furthermore, it can be seen from this survey that city dwellers would be more inclined to use car sharing instead of buying their own car (36%) than respondents from small and medium-sized towns and rural areas (26% and 13% respectively).

It follows from this that the trend of car sharing has a significant tendency to have a negative effect on the demand for cars. However, this does not necessarily have to be to the disadvantage of car dealers. According to Dinsdale et al., car dealers could shift their focus away from the sale of vehicles within the business-to-consumer market towards fleet management, i.e. in cooperation with a car sharing provider, they could manage the purchase, financing, maintenance and repurchase of a mobility provider's vehicle fleet (cf. [21, p. 13]).

3.2.2 The Car Trade of the Future

The effects resulting from societal trends suggest that the car trade must rethink its current tasks and roles in order to be fit for the future. A survey by KPMG (see [22, p. 28]) explains that 48% of the automotive executives surveyed believe it is very likely that the number of physical automotive dealerships will decrease by 30–50% by 2025. Furthermore, 82% of the executives surveyed believe that the only viable option for physical automotive dealerships will be to transform into service factories, used car centers, or focus on an identity management approach where the customer is recognized at every single touch point (see [22, p. 29]).

According to Dinsdale et al., the development of the car trade depends primarily on two critical trends. Firstly, the spread of the sharing economy and secondly, the introduction of autonomous vehicles (cf. [21, p. 4]). Accordingly, *four possible future mobility states result*, which are shown in Fig. 3.1.

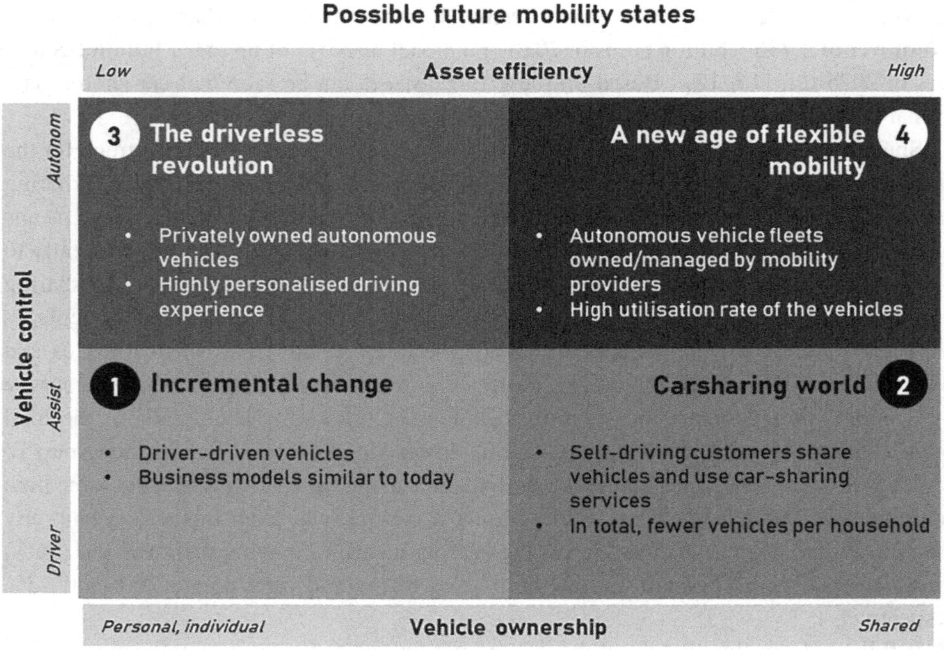

Possible future mobility states

Fig. 3.1 Possible future mobility states. (Own representation based on [21, p. 5])

In the future *mobility state 1 (private, driver-guided cars),* customers will continue to purchase vehicles based on existing product offerings and traditional marketing. While consumers will demand a tailored car buying experience, the key customer value drivers, comparable to today's state, will remain vehicle features and customer experience. The roles of the manufacturer and car dealer in this state will remain the same as today (cf. [21, p. 5]).

In the future *mobility state 2 (sharing of driver-operated cars),* cars are only a means to an end for the customer to get from A to B. The car dealer experience becomes irrelevant. In this future mobility state, car dealers could aim to become mobility service providers, but would have to additionally build new capabilities and business models to do so (cf. [21, p. 6]).

In the future *mobility state 3 (private, autonomous cars),* the automobile has evolved from a simple means of transportation to a personally tailored platform for increased productivity and entertainment. Accordingly, individual configurations and personalization enabled by the car will become the dominant customer value drivers. For car dealers, this consequently means a much higher level of product design, customisation advice and customer support. This also means that car dealerships will become a kind of product experience centers (cf. [21, p. 6]).

In the future *mobility state 4 (sharing of autonomously driving vehicles),* the decisive customer value driver is complete flexibility. However, as in future mobility state 3, personalization, e.g. of vehicle type, media content, and productivity software, remains crucial for the customer (cf. [21, p. 7]). In this future mobility state, car dealers could, as in future mobility state 2, try to act as mobility service providers or concierges. However, this would require an even more profound transformation of capabilities and business model. Accordingly, the most likely transformation is for car dealers to become fleet management services for mobility service providers. In this role, the original car dealer would be responsible for and carry out the purchase, financing and maintenance of a large number of autonomous on-demand vehicles for a mobility service provider (cf. [21, p. 7]).

As the various aspects listed above have shown, general and social trends have a decisive impact on the automotive trade and its future. It can be seen that the automotive trade must rethink its roles and tasks in order to survive in the future.

3.3 Business Processes in the Car Trade

Figure 3.2 depicts the core processes in the car trade. The central processes are vehicle sales, vehicle repair and maintenance, and used car purchasing. The following process descriptions and process models elaborated in detail serve as a starting point for identifying starting points for AI use cases.

3.3.1 Sale of Vehicles

The process of *selling a vehicle* begins with a customer's desire to purchase a vehicle. In order to find a vehicle that suits him and his needs, various types of information gathering are used. For this purpose, the potential customer can go to his "trusted dealer" for advice (cf. [23, p. 3]). Here, the actual process of selling the vehicle would be initiated at the dealership. Furthermore, the potential customer can resort to the widespread use of the Internet to obtain information (cf. [23, p. 3]). Here it is possible to read expert opinions and forum contributions or to visit websites of car manufacturers and to use vehicle configurators. After the customer has found a suitable vehicle for him and his needs from the (over)offer of vehicles, the theoretical result is validated practically, by a test drive. For this, the customer will contact a car dealer (e.g. by phone, by e-mail, by request via an online portal) to make an appointment for the test drive. The car dealer will now enter the lead in the customer relationship management system (CRM system) and mark the

Fig. 3.2 Core processes in the car trade. (Own representation based on interviews with experts)

appointment in a scheduler. Following this, the car dealer will prepare a suitable vehicle for the test drive. Before the vehicle is handed over to the potential customer on the day of the test drive, the driver's license is checked for validity. This is copied and returned to the customer. The customer is then allowed to carry out the test drive. After a successful test drive, the configuration process starts. A prerequisite for this is that the customer requests advice on the appropriate configuration. In order to find a suitable configuration for the customer, an attempt is made to find out his needs and situation. After the successful configuration of the vehicle, the quotation process follows. In this process, an offer is made based on the configuration of the vehicle, the type of purchase (e.g. leasing or purchase), the payment method (cash or financing) and other factors. Depending on the desired payment method, suitable financing partners are searched for and the processes required for this payment method are initiated. Finally, the documents required within the *purchase conclusion* process are signed by the various contracting parties. Subsequently, the vehicle is prepared for handover. This includes ordering the vehicle from the manufacturer. When the vehicle is delivered, it is registered and checked in the workshop. Furthermore, depending on the payment method, the vehicle documents are sent to the financing partner. In a final process, the vehicle is handed over to the customer. The process of selling the vehicle ends with the handover of the keys to the customer.

3.3.2 Vehicle Repair and Maintenance

The *repair process,* shown in Fig. 3.3, is initiated as soon as a customer detects problems with his vehicle and therefore contacts the car dealer to arrange a *repair appointment.* At the repair appointment, the first thing that happens is that the *vehicle is handed over.* This includes the processes of the original vehicle handover, i.e. the handover of the keys, followed by a customer conversation in which the customer relates the symptoms. After the vehicle handover, the *diagnosis* process starts. This process includes the sub-processes of (telemetry) data reading, vehicle inspection and the creation of the diagnosis. After the diagnosis, the actual repair process starts *(carry out repair)*. In this process, the parts that

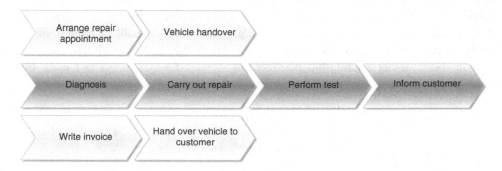

Fig. 3.3 Vehicle repair processes. (Own representation)

may need to be replaced are also ordered, removed and the new parts are reinstalled. This is followed by the *perform test process*. Here it is checked whether the repair was successful. If this is not the case, the process *inform customer* starts, in which the customer is notified. The diagnosis process then starts again, followed by the repair process. If the process *perform test is* positive, that is, the repair was successful, the customer is informed that the repair was successful and that he can pick up his vehicle. After the customer has been informed, the process of *writing the invoice* starts. When the customer picks up their vehicle, it is explained to them what was defective. The invoice is then sent and the vehicle is *handed over*. The processes that occur in the same form for maintenance and repair are shown in light in Figs. 3.3 and 3.4; processes that deviate are shown in dark.

The process of vehicle maintenance (Fig. 3.4) is quite similar to that of vehicle repair, which is why we will not take a closer look at it. It should be noted that the processes of *performing tests* and *informing customers are* omitted in the case of service activities. Furthermore, the processes of *diagnosis* and *repair are* replaced by the processes *read (telemetry) data* and *perform service.*

3.3.3 Used Car Purchase

In principle, the core process of *buying* a *used car* (Fig. 3.5) begins when a car dealer receives a *purchase request.* An appointment is then made with the seller for a closer inspection of the vehicle. This is followed by the *vehicle inspection* process. In this process, the vehicle is technically checked in the workshop, data is read out, the vehicle is examined, and the vehicle documents are checked for correctness. Subsequently, a *test drive is* carried out with the offered vehicle. Subsequently, the process of *evaluating the*

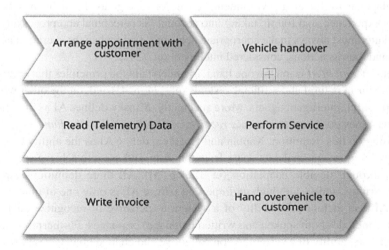

Fig. 3.4 Processes in maintenance. (Own representation)

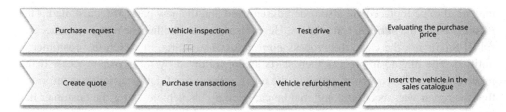

Fig. 3.5 Processes in the purchase of used cars. (Own representation)

purchase price starts. In this process, a purchase price is determined based on the list of defects, which was created in the process of the vehicle inspection, as well as the mileage of the vehicle and other factors. An *offer* is then *made.* If this is to the satisfaction of both parties, the process of the *purchase transaction is* started. In this process, the necessary documents are signed and the vehicle documents as well as the vehicle itself are handed over to the dealer. The dealer deregisters the vehicle. In the following process, the vehicle is prepared in the vehicle *preparation process,* i.e. repairs and a vehicle service are carried out if necessary. After this process, the vehicle is added to the *sales catalog.*

3.4   Artificial Intelligence (AI)

According to Poole and Mackworth, the research field of AI deals in principle with the synthesis and analysis of intelligently acting computer-based agents (cf. [24, p. 3]). Accordingly, a (computerized) agent is intelligent if the following four criteria are met. First, the agent's action must be appropriate for its circumstances and goals, taking into account the short- and long-term consequences of its actions. Second, for the agent to be considered intelligent, he must be adaptable to changing environmental conditions and goals. Third, he must be able to learn from experience, and fourth, taking into account his perceptual and processing limitations, the latter must always make appropriate decisions. Only when these criteria are met is an agent, and thus its actions, considered intelligent (cf. [24, p. 4]).

The central *goal of AI* is, on the one hand, to understand the principles that enable *intelligent behavior* in natural and artificial systems (cf. [24, p. 4]) and, on the other hand, also to develop useful, intelligent agents. More succinctly, Minsky defines AI as the science of making machines *do things that would require intelligence if done by humans* (cf. [25, p. V]). Building on this definition, Kaplan and Haenlein define AI as the ability of a system to correctly interpret external data, learn from that data, and use it to achieve specific goals and tasks through flexible adaptation (cf. [26, p. 3]). All three definitions emphasize machine learning, which, although an important part of AI, is only one of many facets of it. Thus, AI also includes the ability of a system to perceive or recognize data such as speech (this includes both spoken and written language) or images. Furthermore, this field also includes the ability of a system to control, move and change objects based on learned information, be it through a robot or a connected device (robotics).

3.4.1 Subfields of Artificial Intelligence

The field of AI is very multifaceted. In the following, the best-known sub-areas of artificial intelligence are explained, as they are already being used in the economy or in companies.

3.4.1.1 Natural Language Processing (NLP)

Natural language processing (NLP) is the subfield of AI that deals with computers responding not only to formalized programming languages such as Python or Java, but also to natural languages (cf. [27, p. 31]). Natural languages, compared to formalized languages, bear certain challenges that make it difficult for computers to understand and also interpret them. For one thing, the communicated meaning of what is said lies in the semantics of what is said, but semantics is not uniform, as it depends on the respective grammar, cultural imprint or intention, among other things (cf. [27, p. 31]). Furthermore, language changes over time, and language has dialects and accents. Despite grammatical or orthographic errors, what is said or written is understood by human actors. In addition, words can be imprecise. Many words have basically the same meaning but contain certain degrees of nuance (cf. [28, p. 105]).

In order to overcome these challenges of natural languages and thus make them understandable and interpretable for machines, the field of NLP further includes the area of *Natural Language Understanding (NLU)*. The task of the domain NLP is the grammatical and syntactical analysis of texts, which goes beyond the semantic analysis of a text (cf. [29, p. 34]). Functions of the domain NLU on the other hand, have the task of decoding and analyzing the content of texts, i.e. the semantics of texts. It can be seen that certain tasks, such as answering questions, require functions from both the NLP and NLU domains, since it is first necessary to analyze the semantic content of a text before it can be answered syntactically correctly.

NLP can be used in political science, sociology, and psychology to identify and analyze trends, opinions, and common reactions by means of topic modeling and sentiment analysis (see [30, p. 58]). *Chatbots* or dialog agents represent a possible concrete use case of NLP in a company. For example, they could be used in the area of customer service to answer questions and concerns of customers (as chatbot or dialog agent) (cf. [27, p. 32, 30, p. 60]). Furthermore, NLP can be used in the field of reputation monitoring. Here, NLP could be used to perform a sentiment analysis of customer feedback or product reviews, or to improve advertising placements on the Internet through keyword checking (cf. [30, p. 60]).

3.4.1.2 Machine Learning

Machine learning describes the ability of a computer program to learn from examples and experiences, not from previously hard-coded rules (cf. [30, p. 19]). With regard to learning, three types are distinguished: supervised learning, unsupervised learning and reinforcement learning.

In *supervised learning, the* correct answer choices are already known in addition to the actual data (cf. [27, p. 37]). The goal of supervised learning methods is to reveal relationships between input data and output data. Thus, these are about learning from examples (cf. [30, p. 19]). In principle, this form of learning is applied to classifications and regression analyses (cf. [27, p. 37 f.]). Known algorithms of supervised learning are, in addition to K-Nearest Neighbour, Decision Trees, Support Vector Machines and linear regression (cf. [31]).

Learning methods of *unsupervised learning* are supposed to find patterns in the data sets independently and without preceding target values and to form clusters or compress the data based on these patterns (cf. [27, p. 38]). Consequently, the goal of these methods is to discover hidden patterns. A well-known algorithm of unsupervised learning is the k-means algorithm (cf. [31]).

Within *reinforcement learning,* the system should independently find new solution paths (cf. [27, p. 38]). The system must iteratively find an optimal solution path through trial and error. For this purpose, good decisions are encouraged by reward and bad ones are sanctioned by punishment. The system incorporates a variety of environmental influences into its decisions and is also able to react to them. Well-known algorithms of reinforcement learning are Q-learning and Temporal Difference (TD) (cf. [31]).

The use cases for machine learning are almost unlimited, e.g., for making diagnoses in medicine with the help of neural networks or for supporting credit applications (cf. [32, p. 372]). In the automotive industry, machine learning can help improve customer satisfaction, or *predictive maintenance* facilitates maintenance and repair (cf. [28, p. 46 f.]). For the automotive industry, computer vision is still relevant as a subarea of machine learning. Computers identify recurring patterns in images and recognize objects such as vehicle components (cf. [33]).

3.4.2 Economic Impact of Artificial Intelligence

As can be seen in the previous section, AI can be used in many ways and is now omnipresent in everyday economic life (cf. [4, p. 4]). In principle, it is assumed that AI affects the economy in two ways. On the one hand, through productivity increases induced by AI and, on the other hand, through increased consumer demand induced by AI (cf. [4, p. 4, 5, p. 9ff.]).

By using AI, *routine tasks* can be *automated* so that employees can devote more time to tasks with a higher added value. According to Gillham, the use of AI in the value chain can have a positive effect on productivity in various ways (see [5, p. 10]). For example, in the value chain elements of strategy, business model, products, and services, AI-generated simulation of market conditions for creating production forecasts and pricing strategies, or creating digital mock-ups of product features based on historically successful features/user preferences, can reduce the risk of capital expenditure and time, in moving from strategy to implementation. In the areas of marketing, sales and customer service, the use of AI, for example, by using *AI-based chatbots as customer service agents* or via *personalized recommendations of* products and services, can minimize the information asymmetry between customers and companies and adapt communication to the needs of the customer. This increases customer acquisition and customer loyalty.

On the demand side or the consumers, the use of AI has an effect on the one hand through *better product personalization,* increased *product quality* and *time savings* (cf. [5, p. 11ff.]). For example, through the use of AI, on the one hand, consumer preferences can be better identified due to the more detailed data collected about a customer journey, and on the other hand, conclusions can also be better drawn about the factors influencing consumption. In this way, products and services can be tailored more precisely to the customer using AI, based on the optimized data and thus better insights into consumer preferences. On the one hand, this can lead to an increase in the marginal utility of consumption due to product personalization and thus to more consumption. On the other hand, personalized products can increase diversity for the consumer, as previously homogeneous products become more heterogeneous. In terms of product quality, AI can, for example, improve the inherent quality of a product or service through the use of recommendation systems.

3.5   Use Cases of Artificial Intelligence in the Car Trade

Within this section, the possible use cases of AI are presented grouped according to the main service provision processes from Sect. 3.3.

3.5.1   Artificial Intelligence in Vehicle Sales

In the processes of vehicle sales, AI can be used as a *way of optimising customer contact.* This can be done through the use of *chatbots* or dialogue systems. Furthermore, AI can be used to *find a vehicle that is suitable for the* potential customer and his or her needs as well as family, social and financial situation. AI therefore improves the advice and helps to personalise the vehicle and thus also improve the quality of the product or service. For example, dialogue systems or chatbots could be used in the process of obtaining information and advice on the manufacturer's or car dealer's website to *advise the potential customer* and arrange further consultation appointments with a car dealer. Furthermore, dialogue systems or chatbots can be used to *schedule requests for test drives.* AI can also be used in the information gathering and consultation process at the car dealer to find a vehicle and configuration that is adequate for the customer. For this purpose, the potential customer would be asked about various personal characteristics such as age, income, hobbies, gender, marital status, children, occupation, etc. The answers obtained in this way could then be used (for example, in the case of a car) to determine a suitable vehicle and configuration for the customer. The answers received in this way could then be forwarded (anonymously) to a system in which machine learning, with recourse to a central data store of the manufacturer, compares them with the characteristic values of other customers and their existing vehicle type. This data store can be designed as a data lake, which leaves the structured or unstructured data in their original raw format and does without validation, formatting and harmonization (cf. [34]). Alternatively, a central data hub that follows a hub-and-spoke approach to provide unified data for different business requirements is a possible data store (cf. [35]). From the central data store, a vehicle type suitable for the customer and his needs and situation could be found and a reasonable configuration could be suggested to the customer. Furthermore, machine learning could be used in the process of *creating a quote*. Here, a suitable type of ownership (e.g., purchase, leasing, or a mobility subscription) could be determined by the method of supervised learning based on various characteristics of the customer (e.g., annual mileage in km, traveler, commuter, etc.).

3.5.2   Artificial Intelligence in Repair and Maintenance

AI can also be used in a variety of ways in vehicle repair and maintenance processes. For example, dialogue systems or chatbots can be used to arrange a repair appointment. In combination with computer vision, the customer could already send pictures of his vehicle, so that an AI system can already determine possible damage by comparing it with the damage and pictures of other cars during the customer's request for a repair appointment. This is done by using a data lake of the manufacturer, where such pictures and data are stored or collected. Estimated costs can also be determined by evaluating the likely diagnosis and accessing the accident and repair costing system. This is subsequently communicated to the customer [36].

Another possible use of AI, again in the form of dialogue systems or chatbots, can be identified in the process of repair in communication with the customer throughout the entire process. Here, dialogue systems or chatbots can be used to provide information on the progress of repairs as well as the repair process and procedure in response to customer queries.

AI can also be used here to recognize the customer by means of voice identification, i.e. the identification of a person based on the characteristics of his voice (cf. [29, p. 127]). In the actual process of repair, AI can be used in the sub-process of diagnosis. In this sub-process, a diagnosis can be made quickly by using machine learning. For example, the vehicle could be scanned and then an initial diagnosis could be made using computer vision, by comparing it to other damaged vehicles (see [36]). Based on this, by reading the vehicle data, having an employee examine the vehicle and note the findings, and using the data from the warranty request system and the vehicle service booklet, all the crucial data can ultimately be brought together and entered into an AI system so that an accurate diagnosis can be made. It should be noted here that within a warranty enquiry system, among other things, it is stored which repairs have already been carried out and which parts have been replaced.

AI can be used in the vehicle maintenance process in the form of *predictive maintenance*. The core idea of predictive maintenance is the proactive recognition of the need for action for the maintenance or repair of machines and systems before a malfunction has occurred (see [29, p. 217]). For this purpose, relevant data are evaluated in real time in order to determine the optimal time for maintenance or repair. In terms of the car trade and the maintenance process, this would mean that the vehicle data (telemetry data) would be constantly monitored and, when the maintenance case occurs, the respective customer would be informed of the need for a maintenance appointment via a dialogue system or a chatbot. Accordingly, this chatbot has the possibility to access the customer contact data by connecting to the CRM system. Furthermore, the dialog system could already suggest possible free appointments by accessing a scheduling system (workshop scheduling system). By proactively recognizing the need for maintenance, the necessary spare parts can be ordered in time to minimize the downtime of the vehicle (see [29, p. 217]).

3.5.3 Artificial Intelligence in Used Car Purchasing

In the process of buying a used car, a possible use case of AI could be in the process of *checking the vehicle*. In this process, machine learning can be used to check the technical condition of the vehicle. Additionally, the vehicle can be scanned and then computer vision can be used to identify possible damage. In combination with the technical data extracted from the vehicle and the results of the damage analysis obtained from the scanning and via computer vision, it would be possible to determine the condition of the vehicle and whether repairs are required.

Another use case of AI in the area of used car purchasing can be identified in the process of *evaluating the purchase price*. Here, machine learning could be used to determine what an appropriate purchase price is and how long the vehicle would approximately stand until it is sold, based on the extracted vehicle data and with recourse to data from similar used vehicle models from the central data hub or data lake of the manufacturer.

3.5.4 Evaluation of Artificial Intelligence Use Cases

In this section, the previously presented use cases of AI in automotive retailing are evaluated in terms of whether the use cases are already used in the industry or whether they are completely novel. The evaluation of the individual use cases can be seen in Table 3.1.

A system that uses computer vision to assess damage to vehicles is being developed by the company ALTOROS and is to be used in the insurance industry (car insurance) and the car trade (repair) (cf. [37]). Thus, it seems to be a matter of time before such an application becomes widespread in the car trade. Accordingly, it is assumed that these use cases will be integrated into the processes of repair and used car purchase over time.

Table 3.1 Evaluation matrix of AI use cases in the car trade

AI use cases	Processes			
	Vehicle sales	Repair	Maintenance	Used car purchase
Computer vision		2		2
Chatbots, dialogue systems	3	3	3	
Recommendation systems	3			
Repair diagnostic tool		3		
Predictive maintenance			1	

Legend:
1 = already exists and is already in use
2 = is on the rise
3 = novel
Source: Own representation

The use cases of chatbots and dialog systems in the various processes (vehicle sales, repair and maintenance) must be considered in detail. Although chatbots and dialog systems are on the rise (cf. [29, p. 138]), they fulfill different purposes. For example, it can be seen that chatbots or dialogue systems are only offered to a limited extent in the provision of advice/information. Furthermore, according to expert interviews, the majority of car dealerships do not yet use chatbots or dialogue systems in the processes of vehicle sales, repair, maintenance or used car purchasing. It can be concluded from this that the use cases for chatbots or dialogue systems outlined above, also in combination with other AI systems, are "novel" at least for this industry. However, there are already suitable procedures for the economic evaluation of the use of chatbots (cf. [38]).

The use case of AI in the form of machine learning in the process of vehicle sales, in which AI is used, among other things, to gain better insights into the situation and needs of the customer and from this to find a vehicle tailored to his needs and situation in every respect, is reminiscent in its basic idea of recommendation systems. It should be noted that recommendation systems are already used in various industries to better adapt products or services to the customer, e.g. Amazon or Netflix (cf. [39]). As far as can be seen, the use of such a system is not known in the automotive trade, so that this use case is considered "novel".

For the use of machine learning for diagnostics in the repair process, just as in the use case already discussed, no such system is known in the automotive industry, so that this use case is also rated as "novel" here.

The use of AI in the form of *predictive maintenance* is already almost ubiquitous; predictive maintenance is used in the aviation industry (cf. [29, p. 214]) and even in a slight variation (predictive servicing) in the consumer market (cf. [29, p. 220]). It is therefore not surprising that predictive maintenance is being used at least to some extent in the automotive trade. According to interviews with experts, Mercedes-Benz is making use of this in the truck sector. The only question is therefore when this application of AI will also be used for passenger cars.

Overall, it can be stated that some of the developed use cases are already used in the car trade or are at least under development. Others, despite being used in various other industries, are not or rarely used for various reasons. Other use cases are not yet in use. The use of AI in the automotive trade represents great (application) potential and that it is therefore only a matter of time before this potential is also used.

3.6   Conclusion

The social trends affecting the automotive industry mean that potential customers, especially in urban areas, tend to question individual means of transport such as their own vehicle and are more open to collective means of transport or car-sharing services. However, it must be noted that the trend of the sharing economy (car sharing) is still in the development stage and, at least in Germany, does not yet pose an acute threat to the automotive industry or the car trade.

The social trends are also reflected in the possible scenarios of roles and tasks that the automotive trade could take on in the future, such as the role of fleet manager for a mobility service provider. Nevertheless, it can be stated that the possible future tasks and roles of the automotive trade will not be fundamentally different from those of today. Rather, the possible scenarios reflect that the car trade should specialize, for example in fleet management for mobility service providers.

The current core service processes in the car trade are basically multi-layered and strongly characterized by frequent customer contact, especially in the processes of vehicle sales and the processes of vehicle maintenance and repair. In these processes, AI can be used on the one hand to optimize customer contact (e.g. by means of dialog systems or chatbots) and thus bring about an increase in productivity, as employees can devote themselves to more important tasks. On the other hand, AI can be used here to find the customer a vehicle tailored to their needs and situation and in the appropriate ownership type. Thus, by using AI in these processes, the product can be "customized". In the processes of repair and maintenance/service of vehicles, AI can be used above all for faster and more accurate diagnoses and thus accelerate the overall process of repairs and optimize it in a goal-oriented manner. Thus, here too, the use of AI could mean an increase in productivity and customer satisfaction. Further, in the core service process of maintenance, AI can be used in the form of predictive maintenance. Since the possible scenarios of car dealerships in the future imply that car dealers should specialize and since the tasks in the different scenarios are not fundamentally different, with some probability the identified use cases of AI can also be used in the future scenarios. It can be seen that there is also great potential for the application of AI in the automotive trade, which is just waiting to be exploited.

References

1. Hess T (2019) Digitale Transformation strategisch steuern. Springer Fachmedien, Wiesbaden
2. Germany Trade and Invest Gesellschaft für Außenwirtschaft und Standortmarketing mbH (2018) The automotive industry in Germany. https://www.gtai.de/resource/blob/64100/817a53ea339 8a88b83173d5b800123f9/industry-overview-automotive-industry-en-data.pdf. Accessed 18 Feb 2020
3. Parment A (2016) Die Zukunft des Autohandels: Vertrieb und Konsumentenverhalten im Wandel—wie das Auto benutzt, betrachtet und gekauft wird. Springer Fachmedien, Wiesbaden
4. Rao A, Verweij G (2017) Sizing the prize: what's the real value of AI for your business and how can you capitalise? https://www.pwc.com/gx/en/issues/analytics/assets/pwc-ai-analysissizing-the-prize-report.pdf. Accessed 18 Feb 2020

5. Gillham J (2018) The macroeconomic impact of artificial intelligence. https://www.pwc.co.uk/economic-services/assets/macroeconomic-impact-ofai-technical-report-feb-18.pdf. Accessed 18 Feb 2020

6. Winkelhake U (2017) Die digitale Transformation der Automobilindustrie: Treiber—Roadmap—Praxis. Springer, Berlin/Heidelberg

7. Bormann R, Fink P, Holzapfel H et al (2018) Die Zukunft der Deutschen Automobilindustrie—Transformation by Disaster oder by Design. https://www.fes-japan.org/fileadmin/user_upload/14086-20180205.pdf. Accessed 18 Feb 2020

8. Hess T (2019) Digitalisierung. https://www.enzyklopaedie-derwirtschaftsinformatik.de/lexikon/technologien-methoden/Informatik–Grundlagen/digitalisierung. Accessed 19 Feb 2020

9. Pousttchi K (2017) Digitale transformation. https://www.enzyklopaedie-der-wirtschaftsinformatik.de/wienzyklopaedie/lexikon/technologien-methoden/Informatik–Grundlagen/digitalisierung/digitale-transformation/digitaletransformation/. Accessed 19 Feb 2020

10. Hamidian K, Kraijo C (2013) DigITalisierung—Status quo. In: Keuper F, Hamidian K, Verwaayen E et al (eds) Digitalisierung und Innovation: Planung—entstehung—entwicklungsperspektiven. Springer Fachmedien, Wiesbaden, pp 1–23

11. Steinmetz N (2019) Sharing Economy—modelle und Empfehlungen für ein verändertes Konsumverhalten. In: Heinemann G, Gehrckens HM, Täuber T (eds) Handel mit Mehrwert: Digitaler Wandel in Märkten, Geschäftsmodellen und Geschäftssystemen. Springer Fachmedien, Wiesbaden, pp 229–255

12. Bendel O (2019) Sharing economy. https://wirtschaftslexikon.gabler.de/definition/sharing-economy-53876. Accessed 19 Feb 2020

13. Kreutzer RT, Land KH (2017) Digitale Markenführung. Springer Fachmedien, Wiesbaden

14. Schreiner N, Kenning P (2018) Teilen statt Besitzen: Disruption im Rahmen der Sharing Economy. In: Keuper F, Schomann M, Sikora LI et al (eds) Disruption und transformation management: digital leadership—digitales mindset—digitale Strategie. Springer Fachmedien, Wiesbaden, pp 355–379

15. Horizont und Bundesverband CarSharing (2019) Anzahl registrierter Carsharing-Nutzer in Deutschland in den Jahren 2008 bis 2019. https://de.statista.com/statistik/daten/studie/324692/umfrage/carsharing-nutzer-in-deutschland/. Accessed 19 Feb 2020

16. IfD Allensbach (2019) Anzahl der Personen in Deutschland, die Carsharing nutzen oder sich dafür interessieren, in den Jahren 2015 bis 2019 (in Millionen). https://de.statista.com/statistik/daten/studie/257867/umfrage/carsharing-interesse-und-nutzung-in-deutschland/. Accessed 19 Feb 2020

17. Freese C, Schönberg TA (2014) Shared mobility—how new businesses are rewriting the rules of the private transportation game. https://www.rolandberger.com/publications/publication_pdf/roland_berger_tab_shared_mobility_1.pdf. Accessed 19 Feb 2020

18. Frost & Sullivan (2017) Prognostizierter Umsatz für New Mobility Dienste weltweit nach Art der Dienstleistung in den Jahren 2015 und 2025 (in Milliarden US-Dollar). https://de.statista.com/statistik/daten/studie/874212/umfrage/prognose-fuer-den-globalen-umsatz-von-new-mobility-dienste-nach-art/. Accessed 19 Feb 2020

19. Shaheen SA, Cohen AP (2012) Carsharing and personal vehicle services: worldwide market developments and emerging trends. Int J Sustain Transport 7(1):5–34

20. Trendmonitor Deutschland (2018) Können sie sich vorstellen, Carsharing zu nutzen anstatt einen eigenen Pkw anzuschaffen?. https://de.statista.com/statistik/daten/studie/219066/umfrage/carsharing-als-alternative-zum-fahrzeugkauf/. Accessed 19 Feb 2020

21. Dinsdale A, Willigmann P, Corwin S et al (2016). The future of auto retailing: preparing for the evolving mobility ecosystem. https://www2.deloitte.com/content/dam/Deloitte/tr/Documents/consumerbusiness/auto-retailing.pdf. Accessed 19 Feb 2020

22. KPMG (2019) KPMG's global automotive executive survey. https://automotive-institute. kpmg.de/GAES2019/downloads/GAES2019PressConferenceENG_FINAL.PDF. Accessed 19 Feb 2020
23. Kornberg P (2019). Online Lead Management im Automobilhandel 2019. https://i-b-partner. com/wp-content/uploads/2019/10/2019-09-13_RZ-IBP-Whitepaper-OLM-2019.pdf. Accessed 8 Nov 2019
24. Poole D, Mackworth A (2017) Artificial intelligence: foundations of computational agents, 2nd edn. Cambridge University Press, Cambridge
25. Minsky ML (2015) Semantic information processing. MIT Press, Cambridge
26. Kaplan A, Haenlein M (2019) Siri, Siri, in my hand: who's the fairest in the land? On the interpretations, illustrations, and implications of artificial intelligence. Bus Horiz 62(1):15–25
27. Gentsch P (2018) Künstliche Intelligenz für Sales, Marketing und Service: Mit AI und Bots zu einem Algorithmic Business—Konzepte, Technologien und Best Practices, 2nd edn. Springer Fachmedien, Wiesbaden
28. Taulli T (2019) Artificial intelligence basics. Apress, Berkeley
29. Kreutzer RT, Sirrenberg M (2019) Künstliche Intelligenz verstehen: Grundlagen—use-Cases—unternehmenseigene KI-Journey. Springer Fachmedien, Wiesbaden
30. Akerkar R (2019) Artificial intelligence for business. Springer International Publishing, Cham
31. Fumo D (2017) Types of machine learning algorithms you should know. https://towardsdatascience.com/types-of-machine-learning-algorithmsyou-should-know-953a08248861. Accessed 20 Feb 2020
32. Welsch A, Eitle V, Buxmann P (2018) Maschinelles Lernen HMD 55(2):366–382
33. Nord T (2018) Was ist maschinelles Sehen. https://www.lernen-wie-maschinen.ai/ki-pedia/was-ist-maschinelles-sehen/. Accessed 4 März 2020
34. Lubner S, Litzel N (2018) Was ist ein Data Lake?. https://www.bigdata-insider.de/was-ist-ein-data-lake-a-686778/. Accessed 4 März 2020
35. Fagan N (2017) Data lakes, hubs and warehouses. https://neilfaganblog.wordpress. com/2017/07/17/data-lakes-hubs-and-warehouses-fact-vs-fiction/. Accessed 4 März 2020
36. Kwartler T (2017) How AI is changing the way we assess vehicle repair. https://venturebeat. com/2017/02/04/how-ai-is-changing-the-way-we-assess-vehicle-repair/. Accessed 20 Feb 2020
37. Altoros (2019) Car damage recognition|AI in auto insurance. https://www.altoros.com/car-damage-recognition. Accessed 20 Feb 2020
38. Schacker M, Fuchs A (2018) Chatbots im Kundenservice: ein Verfahren zur Kosten-Nutzen-Analyse. Wirtsch Inform Manag 10(8):8–17
39. Wörndl W, Schlichter J (2019) Empfehlungssysteme. https://www.enzyklopaedie-der-wirtschaftsinformatik.de/wi-enzyklopaedie/lexikon/daten-wissen/Business-Intelligence/ Analytische-Informationssysteme–Methoden-der-/empfehlungssysteme/index.html. Accessed 20 Feb 2020

Prof. Dr. Michael Gröschel is a professor at the Faculty of Computer Science at Mannheim University of Applied Sciences. For many years, the business informatics graduate has been involved in research and teaching on topics of business process management and the sensible use of IT in companies in the context of new business models (digital transformation) and the effects on the IT landscape in companies. In addition, he works as a trainer with a focus on business process modeling in BPMN. Twitter: @taxxas

Prof. Dr. Gabriele Roth-Dietrich holds a degree in physics and a doctorate in business administration from the University of Mannheim on process optimization and automation in healthcare. She worked for almost 10 years as a project manager and system analyst in development and product management at SAP SE. After a professorship at Heilbronn University, she has been teaching business informatics at Mannheim University of Applied Sciences since 2011, focusing on enterprise software, workflow management, business intelligence, project management and digital transformation.

Carl-Christian Neundorf is a graduate of the Business and Information Systems course at Mannheim University of Applied Sciences. After graduating in 2020, he worked as an application developer at aubex GmbH in Hockenheim, an IT service provider for the Mercedes-Benz dealer network. Since 2022, Mr. Neundorf has been working as a software developer at solute GmbH, a future-oriented and visionary company in e-commerce. Mr. Neundorf's interests range from AI to digital transformation. He also holds a teaching position at the University of Applied Sciences Mannheim for the subject Python (Faculty of Mechanical Engineering).

Acceptance of AI Systems in Retail

4

Georg Rainer Hofmann and Meike Schumacher

Abstract

This article examines the acceptance of AI systems in retail. To this end, five AI application scenarios are examined in more detail that are already considered realistic today—at least to some extent—or are already being used in practice. The article describes a small series of expert interviews, and aims to provide a synoptic forecast of how the use of AI applications in retail is likely to develop. In this respect, commercial trading companies are given indications of where investments in AI systems could be profitable, as increased acceptance by customers or other stakeholders can be expected.

4.1 Introduction to the Topic

There is currently no generally accepted encyclopaedic definition for the term "Artificial Intelligence—AI". Typically, AI can be seen as a smooth transition from classical automata to systematic algorithmic adaptations—so-called "machine learning". We find on Wikipedia a definition of AI as a "intelligence demonstrated by machines, as opposed to the natural intelligence displayed by animals including humans. Leading AI textbooks define the field as the study of 'intelligent agents': any system that perceives its environment and takes actions that maximize its chance of achieving its goals. Some popular accounts use the term 'artificial intelligence' to describe machines that mimic 'cognitive' functions that humans associate with the human mind, such as 'learning' and

G. R. Hofmann (✉) · M. Schumacher
Information Management Institut IMI, TH, Aschaffenburg, Germany
e-mail: hofmann@th-ab.de

© The Author(s), under exclusive license to Springer Fachmedien Wiesbaden GmbH, part of Springer Nature 2024
T. Barton, C. Müller (eds.), *Artificial intelligence in application*,
https://doi.org/10.1007/978-3-658-43843-2_4

'problem-solving'" [1]. The discourse is polarized: On the one hand, great expectations are formulated for the potential of AI; AI would solve a whole range of technical and everyday problems, increase the quality of life, and generally contribute massively to national economic esteem—see, for exemplary illustration, Fig. 4.1. On the other hand, there are concerns that AI machines will be able to make human labor superfluous, and even dominate and incapacitate humans. In this respect, there are warnings of a loss of control over AI [2].

The 3/2019 issue of the German government's magazine "schwarzrotgold" reflects the German "official" debate and lists popular examples of useful AI applications: Modern cameras can automatically select the best setting for a subject, voice assistants can answer questions, lawn mowers can intelligently find their way across terrain. AI helps overcome physical limitations via intelligent prostheses, and in medicine, AI can be used to develop targeted therapies. AI methods can help reduce traffic jams and accidents via so-called "autonomous driving" and in traffic control. For many years now, AI robots have been taking heavy or monotonous work off the hands of humans. Robots are also helpful in precision work, such as precise welding, drilling and milling, etc.

Computer systems and robots "learn on their own" and can process ever larger amounts of data faster and faster. However, the use of AI methods in retail is still not very present in this popular as well as political discussion.

According to a survey result in the aforementioned issue of the magazine "schwarzrotgold", 63% of consumers support the use of AI. In its study "Artificial Intelligence—Potential

Fig. 4.1 AI has achieved a considerable level of political attention—and is often associated with human behaviour and appearance. (With the kind permission of the Federal Press Office © Bundesregierung/Steffen Kugler)

and Sustainable Change of the Economy in Germany" in autumn 2019, the eco-Verband der Internetwirtschaft e.V. (Association of the Internet Economy) comes to the conclusion that AI will be of great economic importance: "If AI is used across the board, a growth of the gross domestic product of more than 13% by 2025 (compared to 2019) is realistic. This corresponds to a total potential of approx. 488 billion euros. Of this, approx. 330 billion euros (70%) is accounted for by cost savings and approx. 150 billion euros (30%) by revenue potential for all sectors. In absolute terms, the "trade & consumption" and "energy, environment & chemicals" sectors benefit the most, each with just under EUR 100 billion" [3].

Commercial trade (B2B) will also be affected. According to a 2019 study by ibi-Research at the University of Regensburg, the statement "The use of artificial intelligence and learning systems will have a massive impact on the B2B market in the next 5 years" is affirmed by 48% of respondents with high agreement and by 35% of respondents with medium agreement [4].

Often, users are not fully aware of the use of AI. For example, search engines such as Google have been using AI methods for some time to increase the accuracy of their responses. In online retail, AI-based so-called "Proposal Machines" are already being used to provide prospective customers with purchase suggestions that are as tailored to them as possible.

While the survey by "schwarzrotgold" also revealed that 25% of manufacturing companies rely on AI, for example in the area of "predictive maintenance", the situation in retail is still rather unclear. Due to the great economic importance of retail, it is worth taking a closer look at the use of AI methods in retail and their potential and acceptance prospects. The aim is to consistently consider realistic scenarios that directly support the staff involved and the customers and suppliers of the retail sector.

Currently, "AI" is assumed to be one of the core technologies of the digital transformation par excellence. Questions such as "What can AI do?" are answered sweepingly with "almost everything"—a clear indication of ongoing hype in the field. Between hype (the "principle of hope") and skepticism (the "principle of responsibility"), it is important to arrive at a mediating realistic assessment of where a useful and selective use of AI methods seems sensible and where concrete benefits can be seen.

Five scenarios of AI in retail shall be considered in this paper. They are not related to the popular anthropomorphic systems and "nice" robots (as seen in Fig. 4.1). Some of the scenarios are already technologically feasible—in this respect not utopian—and are also already in practical use.

In his speech at the celebration of the 30th anniversary of the German Research Center for Artificial Intelligence (DFKI) in October 2018, August-Wilhelm Scheer [5] points out that "AI" already experienced its first hype years ago. The so-called "expert systems" and "artificial intelligence" already had a Gartner run in the 1980s and are now experiencing a second hype. What are the reasons for the new explosive rise in attention to AI? They are—according to Scheer—higher computing power, bigger data sets, more advanced algorithms and new business models. In the last few decades, around 20 Moore's cycles have doubled the performance of information technology, i.e. it has now increased by a factor of around one million. These developments reinforce each other and thus lead to a new hype.

Companies understandably try to offer their customers innovative products. While large companies can afford to set up and establish AI pilot projects, small and medium-sized enterprises (SMEs) tend to be more cautious because their economic risk is comparatively greater. The media public raves about fully automated factories and super-intelligent computers that are even supposed to be "mentally superior" to humans. This euphoria harbours dangers of misallocation of resources. Already in 2018, Mertens et al. critically explain the degree of novelty of various concepts in the book "Digitalisierung und Industrie 4.0—eine Relativierung" (Digitalization and Industry 4.0—a relativization), using numerous examples to prove the exaggeration of the hype for practical purposes [6].

Whether AI will be successful in the "second rehash" (according to Scheer) depends on whether appropriate applications are found—and accepted by users and society—that justify the investments made. A broad impact of AI is likely to be achieved primarily through consumer-related applications. In the context of this paper, we see applications of AI in retail as very promising. Not "getting into" an application until the success of the technology is considered assured may mean chasing a development. Since IT is developing very dynamically, it means an increased effort if one still wants to catch up with the "first movers".

4.2   Commerce in Transition—From e-Commerce to New Commerce

Since the early days of retailing, consumers, or customers, have been interested in the reliable availability of good quality goods at fair prices. In return, retailers are interested in selling as many goods as possible at the highest possible prices. In principle, this structure has not changed until today—and it will very likely persist in the future. What has changed, however, are the economic and technical conditions under which trade takes place [7].

The question posed in this paper is whether "AI in retail" promises new access for customers and buyers to goods and services.

In brick-and-mortar retail, the expertise of trusted personnel, or merchants, takes center stage. It is the goal of the human staff to recognize the needs of the customers, to meet the needs expressed by them with an appropriate offer (qualitative, quantitative and price). The customer's demand must be satisfied in order to achieve a turnover—which is matched by a corresponding utility value on the customer's side (see Fig. 4.2).

In Internet-based commerce (e-commerce), on the other hand, the focus is on a reliable formal—and therefore automatable—process. The shop system—in essence, it forms a large vending machine—holds an electronic catalog that allows you to find and browse, order, pay, and arrange for delivery.

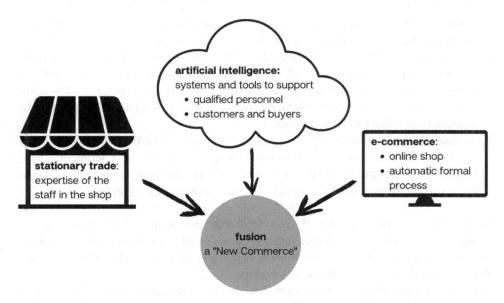

Fig. 4.2 New Commerce as a fusion of brick-and-mortar retail, e-commerce and AI. (© Information Management Institut)

4.3   Acceptance of Systems and Procedures

Every technical advance—also in the Internet economy—necessarily brings with it the question of the acceptance of the technologies, systems and processes developed. With regard to acceptance, a distinction must be made in principle between active and passive acceptance.

Active acceptance means an approval, use and participation, i.e. an active action or at least a positive view on a certain topic. In the use of AI in retail, active acceptance means an acquisition and use of the systems as well as the willingness to bear the corresponding costs.

Passive acceptance, on the other hand, does not describe any active action, but merely a tolerance or toleration of the active acceptance of others. Passive acceptance sometimes even requires corresponding compensation for this tolerance [8]. In the use of AI in commerce, passive acceptance means tolerating the use of the systems by third parties uninvolved in the actual transaction, such as government and administration, associations, consumer advocates. In the socio-political debate, passive acceptance usually has a higher relevance than active acceptance.

Active and passive acceptance rely equally on three –in so far symmetric—factors:

1. Trust in the providers of a good or service: Trust reduces the complexity of the social context and compensates for not knowing about the "exact" functionality and

reliability of the object of acceptance. Trust is acquired in a social context—especially through proven "trust-building" mechanisms of communication.

2. Utility value of the product or service offered: This aspect, of course, includes the price of the product, which is seen in relation to an economic return or a psychosocial and economic utility value. Further aspects are the technical functionality, the statistically measurable reliability of the products or processes, as well as aspects of the pleasure or entertainment value caused by a product.

3. Societal-ethical component: The key question is whether the purchase and use of a good or the use of a service can be considered a justifiable act in terms of ethics. Product properties that are harmful to health or the environment, or violate social customs, such as the prohibition of child labour, are an obstacle to acceptance.

The three factors form a "symmetric tripod" that stabilizes the overall acceptance—if one or two of the factors fail, overall acceptance becomes unbalanced. In particular, the acceptance of systems and services of the highest technical quality can suffer or even be completely undermined by a lack of social acceptance.

When introducing new technologies such as the application of AI in retail, it is not only the active acceptance that directly affects users that is important. If, for example, an item finder is introduced and, in return, the staff that helps customers find products they are looking for in the store is reduced, those who cannot cope with the item finder would neither actively nor passively accept this modernization.

4.4   Five Scenarios of AI in Retailing

As explained at the outset, the term "AI" is not clearly defined and the term "retail" is also characterised by a variety of determinants. However, if the question of the acceptance of AI in retail is to be addressed in the context of a concrete scenario-based study, general definitions to this effect are almost dispensable. When selecting the scenarios, "advertising, disposition and pricing" were identified as the main tasks on the retailer side and "search, selection and ordering" on the customer side. In the following, the five scenarios

1. AI-based information systems
2. AI-based fuzzy matching
3. AI-based plausibility
4. AI-based trade initiation
5. AI-based dynamic pricing

are explained in more detail. Analogous to the AI scenarios and for a better understanding, the respective process is presented as it would occur classically in retail—without the use of AI. For this article, the authors defined these five scenarios as initial acceptance objects for further investigations. In the expert interviews, explicit questions were asked about further scenarios in order to be able to expand the discourse in this regard.

4.4.1 AI-Based Information Systems

So-called "article finder systems" are directly connected to the retailer's product database and, as intelligent information systems for customers, are able to support the sales and consulting process in stationary retail such as a hypermarket. Not only do they provide all the data on all the products on offer in the store, they also use a location description to navigate customers along the most convenient route to the goods' storage location or shelf. To make this navigation even more effective, mobile item finders have been designed for use on smartphones. Thus, the shopping route can always be visualized and displayed to the customer [9].

If, for example, a customer has problems finding a certain product in a consumer or DIY store, it takes several minutes until he has found an employee, can describe his needs to him and he can be guided to the desired shelf. Alternatively, if the customer has the option of entering the desired product into the item finder, they will receive a description, location and pricing information within seconds. This means that the employee does not need to be further involved in the sales and consulting process.

In the store, this could achieve a de facto 100% time saving for the employee on site. The service in the form of an AI information system not only reduces costs, it also increases customer satisfaction through more efficient purchasing options.

4.4.2 AI-Based Fuzzy Matching

Fuzzy matching, also referred to as "fuzzy or error-tolerant search", shows results as reliably and unambiguously as possible, even on the basis of vaguely or incorrectly formulated queries and needs. The application is used when search words are entered incorrectly in technical terms or their exact spellings are not known. In these cases, a "fuzzy" search algorithm can list "matching" search results by shifting or exchanging letters and checking synonyms, including phonetically similar-sounding words. Such search results are displayed, for example, on platforms such as Google with the indication "Did you mean:" [10]. To illustrate the necessity and advantage of a fuzzy matching system in retail, a corresponding simple exemplary sales process in the pharmaceutical industry is outlined:

A customer asks in a pharmacy for a German-named product "Nasensprüh plus". Neither does the pharmacy offer a product with this name, nor have the A pharmacy assistents ever heard of this product. Until both agree on the actual product "Rhinospray plus", several—personnel cost-intensive—minutes pass.

With an AI application, one would now have the possibility to enter the fuzzy term "Nasensprüh plus" into the fuzzy matching system. This could be a pre-competitive application integrated in a pharmacy software or an absolutely independent application "as a service". The search algorithm replaces the letters in such a way that the searched word "Nasensprüh" results in the meaningful term "Nasenspray". With the prefix "Rhino"—meaning "nose"—and the addition of "plus", the AI can ultimately form a link to the product "Rhinospray plus". The search result is displayed to the user within a few seconds. The customer confirms the suggested product. The duration of the process is less than a minute.

With the help of an AI fuzzy matching system, the time required for the sales process could be significantly reduced. Finally, customer satisfaction increases due to fast and successful support, even in the case of fuzzy, inadequately or complexly formulated customer needs.

4.4.3 AI-Based Plausibility

A plausibility check examines order data to determine whether they are credible or plausible. The application scenario of AI plausibility therefore includes, among others, the detection of unrealistic demands, suggestions for a reasonable correction of orders, and compatibility checking [11]. The detection of unrealistic demands includes mistakes that are often made carelessly when ordering from an online store. This is the case, for example, when a product is to be ordered with a quantity of one, but the orderer accidentally entered 11 pieces in the quantity. In a classic, routine handling of the order, the customer would be sent ten pieces too many.

In this scenario, an excessive quantity of the product was ordered due to a careless error. The innovation brought about by the introduction of the AI plausibility system is the checking of orders and purchase orders according to their compatibility (type and quantity), which identifies any errors in the composition or unrealistic requirements. This AI system can be used in all online stores for the ordering process, regardless of whether it is from the clothing, tools, food or any other industry. Accordingly, in this example, in the ordering process, the AI would check for compatibility and logically reasonable quantity with a plausibility check before completing the order. As a result, the system would then issue a warning to the customer about the correctness of the order quantity before completing the order so that the customer can correct it.

In an online shop, for example, we speak of a compatibility check when correlations between products are detected. Possible areas of application here are again the order check or also targeted offers to customers. During the order check, it can also be checked whether the ordered quantity is also "context-sensitive" plausible. If, for example, a customer orders one package of screws and 11 packages of dowels in the appropriate size, it is reasonable to suspect that he is either ordering ten packages of screws too few or ten packages of dowels too many.

With personalized offers, the customer is offered a "plausible" suitable product, often cheaper than it would otherwise be. This would be the case, for example, with a reduced price for screws when purchasing a screwdriver set—with which we see a dovetailing with the scenario of "AI-based trade initiation".

The comparison with the human process is difficult in the plausibility scenario because the processes in which AI would be used make little sense in stationary retail. For example, hardly anyone in stationary retail would accidentally buy 11 packs of a product when they only wanted one.

4.4.4 AI-Based Trade Initiation

Trade initiation is the preparation of the acquisition or sale of an asset. A potential customer enters the dealer's sphere of influence with the aim of concluding a purchase contract. In the foreground of trade initiation is the recognition of a need in the prospective customer that must be satisfied. From the dealer's perspective, it is not enough to simply create a need in order to turn the customer into a buyer. Only when a need for a consumer good arises in the potential buyer does it have relevance for the retailer. In online retail, demand recognition is used to make an offer to the customer with personalized and targeted advertising. Although targeting techniques (keyword, context, CRM, behavioral, retargeting, etc.) are gradually becoming more precise, quite a few customers still receive misplaced advertising that is considered annoying and does not generate any added value for the company despite the high advertising costs [11].

In the following, a simple online retail sales scenario is presented as an analogy example to showcase the effectiveness of AI trade initiation:

A customer wants to order a drilling machine in an online shop and automatically receives the information that customers who bought this product also bought a set of drills. The customer's interest is aroused and he buys both items. Or the interested party receives the information that there is a technical advancement of the drilling machine—the customer then chooses the "better"—but also more expensive—product.

A customer wants to order a book (a novel) in an online shop and automatically receives the information that customers who bought this product have also bought another book, e.g. the sequel to the novel. The customer's interest is aroused—again—and he buys both items.

In order to transfer this scenario to stationary retail, an AI trade initiation is integrated into an AI information system in the form of an item finder. The AI trade initiation system is not only responsible for serious demand recognition, but can also identify new future demands that the customer is currently unaware of based on purchasing behavior and purchases made. Accordingly, the customer can inform himself about offered goods in the article finder or be informed by the AI via a mobile application connected to the point of sale.

4.4.5 AI-Based Dynamic Pricing

Price optimization describes the entirety of all pricing procedures with which the purchasing behavior of customers is influenced in the retailer's favor through dynamic price changes. The desired goal remains the increase of the profit that is generated with a constant offer [12]. Classical case of price optimization is the creation of several differentiated price scenarios for goods such as hotel accommodation, flight bookings or fashion products depending on the time, demand and supply. For fashion items, price reductions are assumed in advance depending on the phase in the product life cycle. This form of pricing is called "markdown optimization" [13].

Another variation is tactical price optimization. In the case of substitute products, ecometric models are used to calculate which goods with high unit yields are not purchased due to an unattractive purchase price. This also includes the non-purchase of more favourable purchase prices. Subsequently, the price of consumer goods with a low unit yield is increased in favour of the retailer, while goods with a high unit yield are reduced in price. Thus, from the buyer's perspective, the products that yield the highest profit for the seller are considered desirable [14].

Price elasticity has a further influence on price formation. Goods with low price elasticity do not show a great impact on price changes in sales. The use of pricing methods can also be considered on a geographical, temporal and individual level. In the geographical observation, a different price is determined not only due to different regional purchasing power, but also as a result of higher expenses incurred in sales (rent, personnel costs: minimum wage in Germany, delivery routes, economies of scale, etc.).

In sales planning, the goods and material requirements of companies could be reliably forecast using corresponding "machine learning algorithms". Factors in the form of demand, trends, weather, sales history, overall market development, market volume and market potential are used for this purpose. The analysis of actual sales ensures that the adaptive algorithms are able to weight the individual factors more and more effectively and thus determine a continuously more accurate demand over time. The advantages for the companies are reflected in better planning of inventories with the help of precise forecasts, increased profits as well as minimization of depreciation and optimal production control [15].

By using AI, not only can the maximum yield per unit be achieved for seasonal products. It is also excellent for perishable foods, the price of which can be optimized over time to avoid losses from discarding products and disposal costs. The potentials of this AI application can not only be profitable, they can also make a positive contribution to the sustainable and conscious use of resources.

4.5   Assessments and Synopsis by Expert Interviews

In the qualitative expert interviews conducted, the focus was on trends and perspectives as well as possible obstacles in the development and establishment of AI applications in retail. In addition, the five identified AI scenarios were presented to the experts in order to subject them to an individual evaluation. The interviewees were selected in order to provide the most comprehensive argumentative coverage of the subject of the discourse. The following six experts participated in the series of interviews:

- Dr. Matias Bauer (KPMG)
- Bernd Bütow (Verband der Vereine Creditreform)
- Dr. Lorenz Determann (REWE)
- Jörg Glaser (ZGV—The Association of Medium-Sized Businesses)
- Prof. Dr. Antonio Krüger (DFKI)
- Klaus Reichenberger (intelligent views)

The main statements and findings from the interviews are summarised below. The names of the interviewees are not mentioned in the following rendering of the results in the sense of an individual assignment and identification of the interviewees.

4.5.1   Assessment of the Importance and Definition of AI

The interviews confirmed: Currently, a common understanding of AI is still missing. The essential characteristic of AI is, on the one hand, that human mental performance is supported and represented by the computer. On the other hand, the term "AI" refers to methods and procedures of self-learning algorithms to make predictions ("predictive analytics"). According to the experts, the systems do not have to "learn" to be AI. Instead of "learning", it is better to speak of "training" or "adapting". True machine "learning"—in its full social complexity—does not yet exist.

The experts see the current AI hype as quite useful—as a "tailwind", so to speak, for stimulating investments in this direction—for further development and research. This hype will presumably also weaken again, but—compared to previous AI hype—it will be more sustainable, as there are now a large number of industrial application possibilities. Occasionally, expectations have to be tempered somewhat, as AI cannot solve every problem. AI can be used when the processes of the application can be formally described. Especially for retail, the AI hype is a challenge, as traditional retail is rather reactive with regard to the generation and evaluation of large and qualitatively sufficient amounts of data required for AI systems. There would be no widely established closed theory ("how to do it") on the use of AI for customers and commerce.

4.5.2   Assessments of the Detectability of AI Applications

According to the unanimous opinion of the experts, AI systems should be recognizable as machines when interacting with humans (especially in dialog systems, such as "chatbots") and should not pretend to be human. An exact and complete imitation of human behavior is not possible anyway, so that a too human impression tends to lead to irritation and problems with acceptance. However, if it is clearly recognizable that a machine is being spoken to, the impression can be "human" (as it is the case with care robots, for example).

4.5.3   Assessment of Barriers

One obstacle, especially for the development and testing of AI applications, is posed by the current data protection legislation in Europe, according to the experts. Innovative companies in Europe are slowed down by the legislation, according to the experts, while competitors in the USA or Asia can operate without these restrictions.

The interviewees expressed the opinion that "common patterns" identified by AI applications should also be viewed critically: People become dull if they are always (only) presented with what they have "always" bought. Showing news is important for the success of advertising. Cross-selling only works if it is intelligent, which in turn depends on "good" customer models.

Individual traders are overprotective of "their" data (a cultural problem) and the data processing systems of the different trading networks are also incompatible (a technological structural problem). This is unlikely to be useful for the development of pre-competitive and cross-company scenarios.

Binding availability and inventory maintenance is important. The data must be qualitatively secured—which is not always achieved to an extent that would be desirable. However, this problem is already known from historic marketplace concepts.

4.5.4   Assessing the Role of Trust and Ethics

According to the experts, many people have a veritable trust problem with the providers and operators of AI systems. Transparency and explainability are therefore very important. In addition, the systems need to be manually controllable—in the event of an emergency.

For the future, an explicitly described ethic on how to deal with AI is important. There must be transparency as to when we are dealing with computers and when we are dealing with humans, and it must also be possible to switch off an "automaton" or to manually bypass its (mal)function.

The experts pointed out that AI could be used to calculate more "results" than can ultimately be used, especially for ethical reasons. Only the information about a customer that is used for a realistic business case should flow into a customer model. The collection and storage of personalized information can also be overridden.

Another problem for trust in the results of AI applications is that the models developed with neural networks, for example, are very difficult—or impossible—to understand. People would ultimately accept AI systems if the benefits were convincing and added value existed. That is why there is a need for serious boundary conditions in the EU area with transparent rules.

4.5.5   Confirmation of the Five AI Scenarios in the Respective Individual Assessment

The experts confirmed that the scenarios presented at the beginning probably cover the AI application areas that are most relevant for retail. In the individual evaluation of the scenarios, the following statements should be highlighted:

AI-Based Information Systems
These systems are already in widespread use. They should be classified as rather less sophisticated AI. Nevertheless, they are relevant. The experts see a lot of potential for the future in AI information systems, especially in the context of omnichannel services and intelligent shopping lists.

AI-Based Fuzzy Matching
Online—for example in Google searches—fuzzy matching already works quite well. In stationary retail, no application scenarios for the customer have been established yet. At the checkout in the store, image recognition is sometimes used, for example, to identify the right articles if they do not have an EAN. A promising application of fuzzy matching can be seen in dialogue systems, such as "intent recognition". Here, the customer tells the system what is important to him when buying a certain product. The AI system draws conclusions from this and ultimately makes a recommendation. The challenge, however, is to identify the product attributes relevant for fuzzy matching and to record them in a high-quality manner.

AI-Based Plausibility
Applications for plausibility checks are always useful when several components of a purchasing portfolio are involved. Here, the experts see a good machine support for the consulting and sales staff in stationary trade.

This application is also relevant for the internal processes of a retail company—as a forecasting method for disposition decisions.

AI-Based Trade Initiation
According to experts, this scenario has the greatest commercial potential, although numerous challenges need to be overcome. One major challenge is identifying the relevant product attributes and maintaining them for a retailer's entire product range. For a targeted initiation of trade, it would also be necessary to recognize different people who use the same end device and possibly search and purchase via the same portals.

AI-Based Dynamic Pricing
Dynamic prices, such as those already commonplace at gas stations or in the travel industry, could theoretically also be a future scenario in stationary retail—even on a customer-specific basis. However, this is also viewed critically by the experts: In traditional stationary retail, pricing is more of a strategic decision and in Germany, people have tended to opt for

largely static prices—especially in the food sector. The price image of a retailer is very important and could be blurred at will by dynamic pricing. What is already being successfully implemented, however, are personalized discounts for customers in a specific product group. The aim here is to open up new product groups for customers. It is also common practice to reduce prices for expiring foodstuffs. It would then become problematic if the mechanisms that lead to the price formation cannot be comprehended by the customer.

Dynamic pricing is based on forecasts, which can be used not only for pricing, but also for quantity or time planning. This is already common practice, but can be made more precise and improved with the help of AI.

4.6   Conclusion and Future Work

Our article comes to the conclusion that the complex technological topic of AI is met positively, at least in larger retail companies, and that they do not trigger any uncertainty or even represent a considerable "fear factor" in this respect.

The interviews gave the impression that there will hardly be any problems with the acceptance of the AI scenarios mentioned, as long as they deliver a comprehensible economic added value. The biggest factors influencing the implementation of AI-supported systems are price and complexity. Opinions are divided, however, on the change in the world of work. While some of the experts expect a shift in professional fields of activity and even the loss of jobs, others do not necessarily see the change as predictable. They see no major impact on the labor market.

Witnn this work, it was not possible to estimate an exact timeframe for when(!) which AI applications will be used on a large scale in which industry. However, it can be seen from the current development that there will be more and more practical applications and mature research in the next few years. For example, one of the interviewees is already using four of the five AI applications described—with thoroughly satisfactory results.

Due to the competitive situation on the market, it is no longer sufficient for retail companies to be active on only one sales channel. As the number of sales channels continues to grow and the trend is towards cross-channel sales, the sales process is becoming more technologically and organizationally complex. Conclusively, this development inevitably ensures a greater scope of AI.

While larger companies and corporations are on the digital march, digitalisation will probably lead to a financial challenge for small and medium-sized trading companies. Precompetitive and also "cooperative" approaches for the use of AI systems—in the sense of "AI as a service"—may need to be considered here.

Meanwhile, the "Great Digitalization" continues—a consolidation of the development is currently not foreseeable. AI methods and systems will play an important role in retail.

At the turn of the millennium, some companies with a large number of customers were of the opinion that all customer dialogues could be completely automated. This was not tenable in the medium term, rather the customers continued to demand a human contact person who understood their concerns. Being able to be reached by telephone proved to be indispensable—and in some cases, the call centers had to be reintroduced. On the other hand, however, the automated customer dialogue can be more efficient. The question therefore arises as to where the right combination of machines and people is to be located, which supports the business processes in retail in the best possible way.

4.7   Best Thanks

The authors would like to thank the students Dejan Bijelic, Daniel Hißlinger, David Spilker, who prepared a seminar paper "Acceptance of application scenarios of AI in retail" in the summer semester 2019 at the TH Aschaffenburg. They contributed decisively to the initial definitions of the scenarios under consideration and evaluated their principle plausibility with a number of regional representatives of commercial trade.

We would also like to thank Andreas Weiss, Emma Wehrwein and Hauke Timmermann from "eco—Verband der Interwirtschaft e. V.", who helped to supervise the above-mentioned seminar paper and provided advice for this publication.

References

1. Wikipedia Künstliche Intelligenz. https://en.wikipedia.org/wiki/Artificial_intelligence . Accessed 7 Apr 2022
2. Hofmann GR (2020) Sind Computer dem Menschen überlegen?—Lesson at the Ringvorlesung "Digitaler Wandel" at TH Aschaffenburg. https://www.mainproject.eu/. Accessed 13 Jan 2020
3. eco-Verband (2019) Neue eco Studie untersucht Wirtschaftspotenziale von Künstlicher Intelligenz: 13 Prozent höheres BIP bis 2025 möglich. https://www.eco.de/presse/neue-eco-studie-untersucht-wirtschaftspotenziale-von-kuenstlicher-intelligenz-13-prozent-hoeheres-bip-bis-2025-moeglich/
4. ibi-Research (2019) B2B-E-Commerce 2020—Status quo, Erfahrungen und Ausblicke. Ergebnisse einer Expertenbefragung von ibi-Research an der Universität Regensburg, Creditreform, eCube und Spryker, Regensburg
5. Scheer AW (2018) Der zweite Aufguss ist stärker. Dinnerspeech at the 30 year anniversary Deutsches Forschungszentrum für Künstliche Intelligenz(DFKI), 18th October 2018, Berlin. https://www.aws-institut.de/im-io/allgemein/der-zweite-aufguss-ist-staerker/?doing_wp_cron=1562572658.9210491180419921875000
6. Peter M, Dina B, Stephan B (2017) Digitalisierung und Industrie 4.0—eine Relativierung. Springer, Wiesbaden
7. Hofmann GR (2011) Quality, pricing and success in electronic retailing—what makes an e-shop successful? World Rev Entrep Manag Sust Dev 7(2):155–173
8. Schumacher M, Hofmann GR (2016) Case-based evidence—Grundlagen und Anwendung: Prognose und Verbesserung der Akzeptanz von Produkten und Projekten. Springer, Wiesbaden
9. Speicher M (o. D.) Artikelfinder. Innovative Retail Laboratory. https://www.innovative-retail.de/index.php?id=86. Accessed 10 Juni 2019

10. IT Wissen (2013) Fuzzy-Suche. https://www.itwissen.info/Fuzzy-Suche-fuzzy-search.html. Accessed 9 June 2019

11. Jansen J (2018) Warum sich die Werbung immer stärker personalisiert?. https://www.faz.net/aktuell/wirtschaft/diginomics/warum-sich-die-werbung-immer-staerker-personalisiert-15788686.html. Accessed 15 June 2019

12. Prudsys, GK Software Group (2019) Dynamische Preisoptimierung. https://prudsys.de/preisoptimierung/. Accessed 15 June 2019

13. Gläß R (2018) Künstliche Intelligenz im Handel 2—Anwendungen. Springer, Wiesbaden

14. Hertel J (2009) Wie taktische Preisoptimierung funktioniert. Absatzwirtschaft, 18. November 2009. https://www.absatzwirtschaft.de/wie-taktische-preisoptimierung-funktioniert-8493/. Accessed 15 June 2019

15. Tidemann M (2018) Der Konkurrenz ein Stück voraus—mit KI-gestützten Absatzprognosen. Blog contribution at Alexander Thamm GmbH, 19. June 2018. https://www.alexanderthamm.com/de/artikel/ki-gestuetzte-absatzprognosen/abgerufen. Accessed 10 June 2019

Georg Rainer Hofmann is a Professor and Director of the Information Management Institute (IMI) at the Technical University of Aschaffenburg. His areas of interest and teaching include, in particular, guiding principles and strategies for data processing and business management, digital transformation, formal systems and philosophy.

Meike Schumacher works in knowledge transfer projects of the European Social Fund (ESF) at the Technical University of Aschaffenburg. Her focus is on methods for determining the context and acceptance of systems and related issues of communication.

Application Potential for Causal Inference in Online Marketing

5

Thomas Barton

*Development of Western science is based on two great achievements:
the invention of the formal logical system (in Euclidean geometry) by
the Greek philosophers and the discovery of the possibility to find out
causal relationships by systematic experiment (Renaissance). Albert
Einstein.*

Abstract

The application of causal diagrams together with the mathematical language of probability theory offers the opportunity to formulate questions of the kind What if? Accessed
and, under certain circumstances, to answer them in a mathematically sound manner. In
this paper, the underlying concept is discussed on the one hand and a possible application in marketing is presented on the other hand, which deals with the effectiveness of
online advertising.

5.1 Introduction

Machine learning today is dominated by statisticians and the belief that you can learn everything from data. This data-centric philosophy is limited

Judea Pearl [1]

Machine learning and neural networks are very present due to spectacular results. There is
a rapidly increasing number of applications that make use of their methods. There is a veritable euphoria that by analyzing very extensive data sets with machine learning or neural
network methods, it should be possible to solve almost any problem. As so-called

T. Barton (✉)
Department of Computer Science, Worms University of Applied Sciences, Worms, Germany
e-mail: barton@hs-worms.de

black-box AI, these methods have the disadvantage that an explanation for the occurrence of a result is not possible or only possible to a limited extent (see for example [2]). In comparison, in the natural sciences and especially in physics it is not only self-evident but even indispensable to explain scientific phenomena on the basis of cause and effect. The mathematical methods of statistical inference offer the opportunity to introduce causal thinking in the form of artificial intelligence methods even in scientific disciplines in which explanations based on cause and effect have not previously been considered indispensable. An example of this is the subject area of marketing.

5.2 Causal Diagrams and Do Operator

> You need to have causal assumptions before you can get causal conclusions, which you cannot get from data alone

Judea Pearl [1]

Figure 5.1 shows an example of a causal diagram. This diagram shows a model for the causal relationship between the acquired University education E_d, the available experience E_x and the resulting salary S of employees. The causal diagram is associated with a structural causal model (SCM). Here, University education E_d, experience E_x and salary S are endogenous variables of the model and depend on the other variables of the model. The variable U_s is an exogenous variable. It depends on factors that are external to the model. Further component of a structural causal model is a set of functions f that assigns to each value of a variable a value that depends on the other variables of the model (an exact definition for SCM can be found in [3]). In the example presented, the function f_S shows the functional relationship between the salary S and the University education E_d, the experience E_x, and the external variable U_s:

$$S = f_S\left(E_d, E_x, U_S\right) \tag{5.1}$$

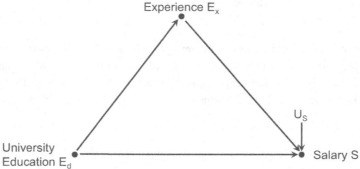

Fig. 5.1 Example of a causal diagram on employee salary

The diagram in Fig. 5.1 shows University education as the direct cause of an employee's salary. The following values are applied for University education: $E_d = 0$ (no college degree), $E_d = 1$ (bachelor's degree), $E_d = 2$ (master's degree), and $E_d = 3$ (doctorate). University education is also considered as a source of experience in this model. Experience E_x is expressed in years of work experience. In this model, an employee's experience is considered to be the cause of an employee's achieved annual salary S, as is higher education. The above example is adapted from [4].

In recent decades, it has become popular to use probability theory as a mathematical language to describe and model causal relationships in various scientific disciplines [3]. For example, if we consider an employee's experience E_x as a variable and the number 20 is its value (which corresponds to years of work experience), $P(E = 20)$ is the probability that an employee has 20 years of work experience. For the conditional probability $P(S|E_x)$ of S under the condition E_x, Bayes' theorem holds according to Eq. 5.2:

$$P\left(S|E_x\right) = \frac{P\left(E_x|S\right) \bullet P\left(S\right)}{P\left(E_x\right)} \tag{5.2}$$

For the considerations on the so-called do operator, we start from the scenario given by the causal diagram in Fig. 5.2. In this model, both variable X and variable Z are considered to be the cause of variable Y. Furthermore, a causal relationship between variables Z and X is also assumed.

The expression $P(y|do(x), z)$ represents the probability that the event Y = y occurs under the conditions that on the one hand the variable X is set to the value x by a so-called intervention and on the other hand the variable Z takes the value z [5]. The intervention of assigning a value to the variable X is described with the help of the operator do(x). In this way, it is possible to ask the following question: What if the variable X were to take the value x?

Work by and around Judea Pearl in particular has led to the possibility of analytically computing the expression $P(y|do(x), z)$ for the example shown in Fig. 5.2 [3]:

$$P\left(y|do\left(x\right), z\right) = \sum_z P\left(y|x, z\right) P\left(z\right) \tag{5.3}$$

Fig. 5.2 Causal diagram for three variables X, Y and Z

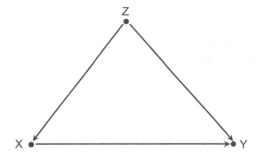

5.3   Application Potentials in Online Marketing

The market for online advertising is extremely attractive with a turnover of € 8 billion in Germany alone [6]. While, according to one study, Internet users perceive online advertising always (13%) and sometimes (59%), 52% of respondents rarely click on it and 28% of Internet users never do [7]. How likely is it that an Internet user will click on a particular online advertisement? With the help of the do operator, the question of how effective an advertising campaign is can be expressed mathematically, provided that a causal model is available. The effectiveness of an advertising campaign should be considered on the basis of the causal diagram shown in Fig. 5.3 (see also [8]).

Here it is assumed that online advertising, represented by variable X, is a cause of a conversion, where the conversion is represented by variable Y. The conversion can be, for example, a click, an inquiry, or an order. A conversion can be, for example, a click, an inquiry, a registration or an order. Another causal relationship is between customer and conversion. Here, a customer is described by various characteristics and influencing variables such as age, gender and income. A causal relationship is also assumed between customer and online advertising. The question of how effective an advertising campaign is can be determined by first checking whether the probability $P(y|do(x), z)$ can be analytically determined on the basis of the causal diagram, and if so, its value can be determined on the basis of data to be collected. For the model at hand, the probability can be determined as follows:

$$P\big(y|do\big(x\big)z\big) = \sum_z P\big(y|x,z\big)P\big(z\big) \tag{5.4}$$

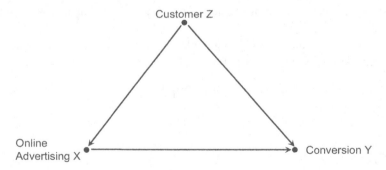

Fig. 5.3 Causal diagram of the effect relationship between customer, online advertising and conversion

5.4   Conclusion

With the help of causal diagrams and Bayesian statistics, questions about the effectiveness of an advertising campaign in the field of online marketing can be formulated mathematically and quantified on this basis. The development of a suitable causal model and its application in concrete projects offer interesting perspectives for the use of artificial intelligence in marketing and sales.

References

1. Ford M (2018) Architects of intelligence: the truth about AI from the people building it. Packt Publishing, Birmingham
2. Lämmel U, Cleve C (2020) Künstliche Intelligenz, 5th Edition, Hanser, München
3. Pearl J (2009) Causality: models, reasoning, and inference, 2nd Edition, Cambridge University Press, New York
4. Pearl J, Mackenzie D (2018) The book of why: the new science of cause and effect. Basic Books, New York
5. Pearl J (2019) The seven tools of causal reasoning with reflections on machine learning. Commun ACM 62:54–60
6. PwC (2019) German entertainment and media outlook: 2019–2023. https://www.pwc.de/de/technologie-medien-und-telekommunikation/german-entertainment-and-media-outlook-2019-2023.html. Accessed 7 May 2024
7. PwC (2019) E-Privacy 2019. Ergebnisse: Bevölkerungsbefragung zu personalisierter Werbung. https://www.pwc.de/de/pressemitteilungen/2019/personalisierte-online-werbung-erreicht-die-nutzer.html. Accessed 7 July 2022
8. Li A, Pearl J (2019) Unit selection based on counterfactual logic. In: Proceedings of the twenty-eighth international joint conference on artificial intelligence. S. 1793–1799. https://doi.org/10.24963/ijcai.2019/248

Thomas Barton studied physics at the TU Kaiserslautern with minors in computer science and mathematics. He obtained his doctorate at the intersection of physics and medicine in the research group of Prof. Dr. Wolfgang Demtröder at the Department of Physics at RPTU Kaiserslautern-Landau. In the course of this work, he gained relevant experience in the software-supported control of experiments and in the analysis of large amounts of data. This expertise led him to SAP SE, where he worked for 10 years with a focus on application development, also consulting, training and project management. Since 2006, he has been working at Worms University of Applied Sciences as a professor of computer science with a focus on business informatics. His work focuses on the development of business applications, e-business and data science. He is the author and editor of numerous publications. In addition, he is active in various committees and expert panels. He is also the spokesperson for the GI Advisory Board for Universities of Applied Sciences.

Influence of Artificial Intelligence on Customer Journeys Using the Example of Intelligent Parking

6

Dominik Schneider, Frank Wisselink, Nikolai Nölle, and Christian Czarnecki

Abstract

New applications of Artificial Intelligence (AI) are increasingly emerging in the consumer market. More and more, devices and services that communicate autonomously over the internet are also entering the market. This can enhance these devices and services with novel AI-based services. Such services can influence the way customers make commercial decisions, significantly changing the customer experience. The impact of AI on commercial interactions has not been studied widely. Based on a framework that provides an initial overview of the effects of AI on commercial interactions, this chapter analyzes the influence of AI on customer journeys using the specific use case of intelligent parking. The insights gained from this can be used in practice as a basis for understanding the potential of AI and implementing it in the design of one's own customer journeys.

D. Schneider (✉) · F. Wisselink · N. Nölle
Deutsche Telekom Gruppe, Bonn, Germany
e-mail: Dominik.Schneider02@telekom.de; Frank.wisselink@t-systems.com;
Nikolai.Noelle@telekom.de

C. Czarnecki
FH Aachen—University of Applied Sciences, Aachen, Germany
e-mail: czarnecki@fh-aachen.de

T. Barton, C. Müller (eds.), *Artificial intelligence in application*,
https://doi.org/10.1007/978-3-658-43843-2_6

6.1   How Does Artificial Intelligence Influence Commercial Decisions?

For some years now, competitive pressure has been increasing in many industries [1, 2], which is why companies need to differentiate themselves from the competition. At the same time, differentiation via prices, product ranges and unique technologies is increasingly losing its importance as a strategic competitive factor [3]. Companies have realized that creating a strong customer experience (CX) will determine future competitiveness and the chance to lead in the market [3, 4]. Customer experience management (CEM) thus takes on a new significance in companies [4–6] and the design of positive customer experiences becomes a strategically important competitive differentiator [4, 7, 8].

CEM generally encompasses the entire strategic management process for shaping a company's CX [9]. CX refers to the sum of all experiences and associated emotions that arise in the interaction between customers and systems or products and services of a company [3]. Accordingly, CX is based on two levels: (1) the emotional level and (2) the process level [4, 9, 10]. The goal of CEM is to purposefully design a holistic consistent customer experience [5] based on emotional level and process level. Since the emotional value of a CX is difficult to infer and highly dependent on the operational design of individual touchpoints, the main focus of this chapter is on the process level. For the design of a CX on the process level, customer journeys are a relevant means of CEM. Customer journeys map the interaction of a customer with a company, a brand or its products through different phases and touchpoints [11]. Compared to CX, customer journeys are more narrowly focused on a customer's buying process [12, 13], which generalized consists of a pre-purchase, purchase, and post-purchase phase [5]. New technological developments such as the Internet of Things (IoT) and Artificial Intelligence (AI) have the potential to significantly change the design of customer journeys, revolutionize CEM, and thus influence commercial decisions [14].

With the IoT, the number of networked sensors, actuators, and traditional objects will greatly increase [15, 16]. As a result, many devices will become capable of communication that previously were not able to communicate [15]. New types of digital touchpoints are emerging that can be integrated into customer journeys and used to communicate CX [14]. However, in the context of CEM, IoT is considered as a technical enabler for AI. AI is the property of an IT system to behave in a human-like manner [17]. The basic idea of AI research is to consider cognitive processes as processes of data processing [18]. Perceiving, understanding, acting, and learning are four core capabilities that enable AI to exhibit human-like behaviors [17]. In terms of designing and evolving customer journeys, AI can help in gaining improved knowledge about customer needs from data collected through digital touchpoints. Based on this, AI can be used to enable automated actions for the optimal, customer-needs-oriented delivery of CX.

Against the background of these new possibilities, a growing interest in the design and further development of customer journeys as well as in AI has developed in the environment of CEM. Research on the influence of AI on customer journeys, on the other hand, is still in its early stages [19]. Especially at the process level, the combined consideration of AI and customer journeys represents a gap in research [19, 20]. To open up this novel research area, Schneider et al. propose a framework [19] that provides a first overview of the effects of AI on commercial interactions. How these effects are derived from so-called patterns and factors identified in numerous customer journey comparisons is the subject of this chapter.

After a motivation for the topic, Sect. 6.2 discusses the state of research. In Sect. 6.3, the procedure for deriving effects from patterns and factors is demonstrated using the concrete example of a customer journey comparison for intelligent parking, before the results of customer journey comparisons of numerous scenarios are generalised and discussed in Sect. 6.4. The chapter ends with a conclusion and outlook in Sect. 6.5.

6.2   State of Research on the Influence of AI on Customer Journeys

Research on the impact of AI on customer journeys, especially on the process level, is in its early stages [19, 20]. To explore this novel research area, Schneider et al. propose a framework [19] that provides a first overview of the effects of AI on commercial interactions. The framework is the result of a seven-step case study and design-oriented approach [19]. The authors first select a total of seven AI use cases and one matching non-AI use case for each [19]. The interrelated consideration of an AI and conventional use case pursuing the same or similar goal is defined by Schneider et al. as a scenario [19]. The seven selected scenarios include intelligent parking (S1), television control by voice assistant (S2), autonomous shopping (S3), online shopping with purchase recommendations (S4), shopping in a department store with intelligent digital assistant (S5), shopping by voice command (S6), voice identification for telephone customer service (S7), and the corresponding equivalent use cases without AI [19]. Customer journeys are modeled for each AI-based and conventional use case [19]. A comparison of the customer journeys is followed by the documentation of the differences per scenario [19]. The documented differences of all scenarios are also compared to identify intersections [19]. As a result, the authors propose a framework that structures the effects of AI on commercial interactions (Fig. 6.1) [19].

This framework is structured according to three categories (1) improved knowledge for customers, (2) process agility, and (3) other effects [19]. In each category, several effects have been identified that result from combinations of so-called factors [19]. These factors are changes that AI causes in different phases of a customer journey (awareness, consideration, purchase, usage, and retention) and that occur repeatedly in numerous scenarios [19]. All these factors are derived from different combinations of patterns [19]. Patterns are changes that AI causes in customer activities or customer sub-processes or touchpoints and that occur repeatedly in different phases of a customer journey as well as in numerous scenarios [19]. In the following Sect. 6.3, the example of intelligent parking is used to demonstrate how patterns, factors and effects are derived, building on Schneider et al. [19].

	Awareness	Consideration	Purchase	Usage	Retention
Improved Knowledge for Customers	• **E1**: More needs-oriented support of awareness creation	• **E3**: More demand-oriented formation of the purchase intention	• **E5**: Need-based purchase decision	• **PE7**: Information-based extension of the product or service benefit	• **PE9**: Optimization of customer service
Process Agility	• **E2**: Faster and more convenient awareness creation	• **E4**: Faster and more convenient formation of the purchase intention	• **E6**: Faster and more convenient purchase decision	• **E8**: Focus on the actual usage with less process-related distraction	• **E10**: Strengthening of the customer dialogue
Other Effects		• **E11**: Increased Discretion • **PE12**: Permanent Service availability • **PE15**: Versatile voice support • **PE16**: Individual, digital consulting in the physical world • **PE18**: Access to ervices independent of knowledge and ownership • **PE19**: Increased Security	• **E11**: Increase Discretion • **PE12**: Permanent Service availability • **PE14**: Satisfaction of other needs • **PE15**: Versatile language support • **PE16**: Individual, digital consulting in the physical world		• **E13**: Customer trust • **PE17**: Loyal customer relationship

Fig. 6.1 Proposed framework: Characteristic effects of AI on commercial interactions. (cf. [19])

6.3 Influence of AI on Customer Journeys Using the Example of Intelligent Parking

Intelligent parking is a scenario in which the goal is to find a parking space for a vehicle at a foreign destination and to pay for the required parking time. In [19], Schneider et al. describe and compare customer journeys for a traditional use case and an AI use case of this scenario. The conventional use case (Fig. 6.2) focuses on the classical search for a parking space without AI and the payment of the parking time with a parking ticket.

The AI use case addresses the search for a parking space and payment of parking time using an intelligent parking solution like eParkomat[1] mobile app [21], as shown in Fig. 6.3.

In the following, the differences between the two customer journeys must be identified and documented. To do this, a deviation analysis is first performed between the customer journeys for the AI use case and the conventional use case. The goal of the deviation analysis is to identify potential patterns. Patterns are defined as changes that AI causes in customer activities or customer sub-processes or touchpoints and that occur repeatedly in different phases of a customer journey as well as in numerous scenarios [19]. The definition of the scenario overarching patterns is done in Sect. 6.4.1, because only after the

---

[1] Note: This article is based on research carried out in 2019. At that time 'eParkomat' was one of the first intelligent parking solutions based in Czeck Republic.

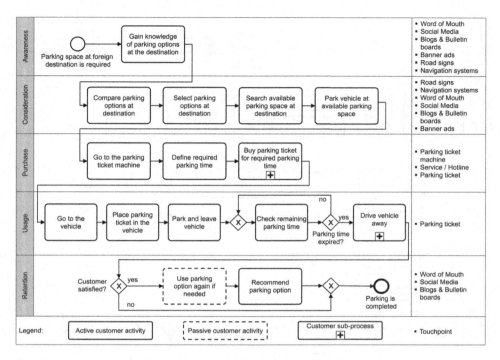

Fig. 6.2 Classic parking with parking ticket without extension of parking time. (cf. [19])

generalization of the customer journey comparison results of several scenarios an overview of potential patterns from different scenarios is given.

6.3.1   Identified Patterns Using the Example of Intelligent Parking

The potential patterns identified in the intelligent parking scenario are listed in Table 6.1. In the following, these potential patterns and their appearance in the customer journeys (Figs. 6.2 and 6.3) are discussed. To refer to the potential patterns, their numbers (see column "No." in Tab. 6.1) are used. For clarity, the phase names in the column "Appearance in phases" in Table 6.1 and in other tables are abbreviated.[2]

When comparing the customer journeys, the first thing that stands out is that existing customer activities or customer sub-processes are eliminated through the use of an intelligent parking solution (PM1). For example, the customer no longer has to go to the parking ticket machine in the purchase phase or back to the vehicle in the usage phase and place the parking ticket in the vehicle. Also, a postponement of a customer activity (PM2)

---

[2] In the following tables, the phase designations in the column "Appearance in phases" are abbreviated as follows: Awareness (A), Consideration (C), Purchase (P), Usage (U), Retention (R).

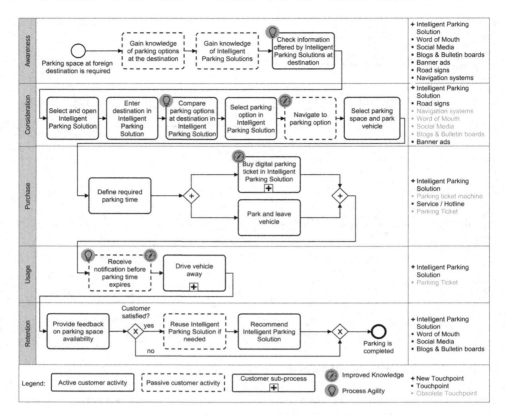

Fig. 6.3 Parking with an intelligent parking solution without extending the parking time. (cf. [19])

Table 6.1 Potential patterns in the parking scenario

No.	Designation of potential patterns	Appearance in phases
PM1	Elimination of existing customer activities/sub-processes	P, U
PM2	Postponement of existing customer activities/sub-processes	P, U
PM3	Addition of new customer activities/sub-processes	A, C, R
PM4	Active-passive change for existing customer activities/ sub-processes	A, C, U
PM5	Automation of customer activities/sub-processes	U
PM6	Parallelization of customer activities/sub-processes	P
PM7	Addition of transparency-creating information	A, C, P
PM8	Addition of site-specific information	A, C, P, U
PM9	Addition of time-specific information	A, C, P, U
PM10	Status change from original to obsolete touchpoints	C, P, U
PM11	Addition of new touchpoints	A, C, P, U, R

from the usage phase to the purchase phase can be observed. This is due to the fact that in the AI use case, the customer can already move away from the vehicle during the purchase phase while booking a digital parking ticket regardless of location. In the conventional use case, this customer activity is only possible in the usage phase. In addition, new customer activities are added in the awareness, consideration, and retention phases of the AI use case (PM3). Examples are the customer activities "Select and open intelligent parking solution" and "Provide feedback on parking space availability".

Furthermore, an active-passive change (PM4) can be observed in three customer activities with regard to the role of the customer. While in the conventional use case, for example, the customer actively acquires knowledge about parking options at the destination, in the AI use case the customer tends to perceive information about parking options at the destination only unconsciously and thus acquires knowledge passively. Another example is the search for a parking space in the consideration phase. In the conventional use case, the customer actively searches for a free parking space at the destination. With the help of an intelligent parking solution, the customer can be navigated to a previously selected parking area and assume a more passive role in the search for a parking space. The fifth potential pattern is the automation of customer activities or sub-processes (PM5). This potential pattern can be seen in the usage phase, where the customer automatically receives a reminder from the intelligent parking solutionbefore the expiration of the booked parking time. This eliminates the need for a manual check of the remaining parking time as in the conventional use case. Another potential pattern is the parallelization of customer activities or sub-processes (PM6). In the purchase phase, the customer can book a digital parking ticket while walking away from the parked vehicle.

Other potential patterns are the addition of transparency-creating (PM7), site-specific (PM8) or time-specific (PM9) information. For example, information about the parking situation at the destination are already provided to the customer in the awareness phase, as he checks the information offer of different intelligent parking solutions at the destination. In the consideration phase, the customer makes extensive use of the information provided by the intelligent parking solution, which creates transparency about the parking situation at specific times around the destination and is not available in the conventional use case. This information is also available in the purchase phase and is a basis for the customer's purchase decision. Another example is the time-specific information played out in the usage phase to remind the customer that the parking time is about to expire. With the reminder, the customer can also receive site-specific information about the location of the parked vehicle.

Potential patterns can also be identified based on the touchpoints. For example, in the consideration, purchase, and usage phases, the status of some original touchpoints changes to obsolete touchpoints (PM10). This means that some touchpoints are replaced by AI or are no longer relevant due to AI, such as the parking ticket machine or parking ticket in the purchase phase. The physical existence of original touchpoints is usually not affected when the status changes to obsolete contact points. For example, parking meters and parking tickets still exist. However, they are no longer relevant in the customer journey due to the use of AI. In addition, a new touchpoint is added with the intelligent parking solutions in the awareness phase or a selected parking solution in all other phases (PM11).

6.3.2 Identified Factors Using the Example of Intelligent Parking

With the identified potential patterns, the first differences between the customer journeys are documented. Building on this, the next step is to identify so-called factors. Factors are derived from different combinations of patterns and are changes that AI causes in different phases of a customer journey and that occur repeatedly in numerous scenarios [19]. The definition of factors across scenarios is given in Sect. 6.4.2. The potential factors derived in the intelligent parking scenario are listed in Table 6.2 and are explained in detail below. The numbers of the potential factors (see column "No." in Table 6.2) are used to refer to the potential factors.

The first identifiable potential factor is called phase narrowing (PF1). Phase narrowing means that the number of sequential customer activities or sub-processes in a phase is reduced. This is the case, for example, when more existing customer activities or sub-processes in a phase are eliminated (PM1) or moved to another phase (PM2) than new customer activities or customer sub-processes are added (PM3). For example, while the usage phase in Fig. 6.2 consists of five customer activities or sub-processes, it only includes two customer activities or sub-processes in Fig. 6.3. The reason for this is that two existing customer activities ("Go to the vehicle" and "Place parking ticket in vehicle") are omitted and one customer activity ("Park and leave vehicle") is moved to the Purchase phase. Other pattern combinations are also conceivable. For example, a parallelization of customer activities or sub-processes (PM6) can lead to a phase narrowing if more customer activities or sub-processes are parallelized than new customer activities or sub-processes are added (PM3) or moved from another phase to the corresponding phase (PM2). This combination of patterns can be observed, for example, in the purchase phase in Fig. 6.3.

In contrast to a phase narrowing, a phase extension (PF2) means that the number of sequential customer activities or sub-processes in a phase increases. A phase extension occurs, for example, when more new customer activities or sub-processes are added to a

Table 6.2 Potential factors in the parking scenario

No.	Designation of potential factors	Appearance in phases
PF1	Phase narrowing	P, U
PF2	Phase extension	A, C, R
PF3	Reduction of throughput times	A, C, P
PF4	Extension of throughput times	R
PF5	Reduction of manual customer effort	A, C, P, U
PF6	Reduction of required customer knowledge	A, C, P, U
PF7	Improvement of situational knowledge	A, C, P, U
PF8	Focusing of touchpoints	A, C, P, U, R

phase (PM3) than existing customer activities or sub-processes are eliminated (PM1), postponed to another phase (PM2), or parallelized (PM6). Other pattern combinations are also conceivable for deriving this potential factor. For example, a phase extension can be seen in the consideration phase in Fig. 6.3, which includes a total of six customer activities, whereas the consideration phase in Fig. 6.2 consists of only four customer activities. One reason for this phase expansion is that two new customer activities ("Select and open intelligent parking solution" and "Enter destination in inteligent parking solution") are added. Phase extensions also occur in the awareness and retention phases.

Another potential factor is the reduction of throughput times (PF3). This potential factor means that a customer needs less time to execute the customer activities or sub-processes from the beginning to the end of a phase. A reduction in throughput time can occur, for example, if more time is reduced in a phase by eliminating (PM1) or parallelizing (PM6) customer activities or sub-processes than time is reduced by new (PM3) or postponed (PM2) customer activities or sub-processes in the corresponding phase. For example, a reduction in throughput time can be assumed in the purchase phase in Fig. 6.3 compared to the purchase phase in Fig. 6.2. A reduction of throughput times can also result from active-passive changes in customer activities or sub-processes (PM4), provided that more time is reduced than the time required by newly added (PM3) or postponed (PM2) customer activities or sub-processes. An example of this is the consideration phase in Fig. 6.3, in which an active-passive change can be observed in the search for a parking space. While in the conventional use case the customer actively searches for a free parking space at the destination, with an intelligent parking solution the customer is navigated to a previously selected parking area with predicted available parking spaces. Although new customer activities are also added in the consideration phase with a small amount of time (PM3), navigating to a parking area with available parking spaces can conceivably reduce the time required by several minutes [22]. A reduction in throughput time can also be observed in the awareness phase.

The counterpart to the reduction is the extension of throughput times (PF4). This potential factor means that a customer needs more time to execute the customer activities or sub-processes from the beginning to the end of a phase. An extension of throughput time occurs, for example, when new customer activities or sub-processes are added (PM3) or postponed to the corresponding phase (PM2), and more time is required than is reduced by the elimination (PM1) or parallelization (PM6) of customer activities or sub-processes. An example of an extension of the throughput time is the retention phase, in which a new customer activity ("Provide feedback on parking space availability") is added by the intelligent parking solution, which can lead to an extension of the throughput time.

In addition, the reduction of manual customer effort (PF5) can be derived from different combinations of patterns. This potential factor means that a customer has to expend less manual effort to execute the customer activities or sub-processes from the beginning to the end of a phase. Manual customer effort has no direct relation to the time dimension and is thus clearly distinguished from the time reduction considered earlier in the context of shortening throughput times (PF3). A reduction in manual customer effort can occur in a phase, for example, by eliminating (PM1), postponing (PM2), or automating (PM5) customer activities or sub-processes that require active customer action. The prerequisite for a reduction in manual customer effort is that more manual effort is reduced in a phase than is created by new customer activities or sub-processes that are added (PM3) or postponded (PM2) to the corresponding phase. The example of the usage phase can be used to illustrate this potential factor. When parking with a parking ticket, there are a total of five customer activities or sub-processes in the usage phase that require manual customer effort. An intelligent parking solution initially eliminates two customer activities ("Go to the vehicle" and "Place parking ticket in vehicle"). In addition, one customer activity is moved to the purchase phase ("Park and leave vehicle") and another customer activity is automated ("Check remaining parking time"). Consequently, only the customer sub-process "Drive vehicle away" remains in the usage phase, which requires manual customer effort. Furthermore, manual customer effort can be reduced e.g. by an active-passive change (PM4). For example, in Fig. 6.3, a reduction of customer effort can be observed in the consideration phase. Since the customer no longer has to actively search for a free parking space at the destination, but can instead use the intelligent parking solution to navigate to a previously selected parking area with predicted available parking spaces, a significant reduction in search effort is likely. Manual customer effort is also reduced by the intelligent parking solution in the awareness and purchase phase.

The reduction of required customer knowledge (PF6) is another potential factor. This means that the passage through a phase is simplified for the customer. Simplification for the customer can occur, for example, by reducing the number of touchpoints requiring customer knowledge in a phase (PM10), provided that more customer knowledge is not required overall through the addition of new touchpoints (PM11). Simplification through changes in touchpoints can be observed, for example, in the consideration phase. There, original touchpoints such as navigation systems, word-of-mouth communication, social media, blogs and bulletin boards for comparison and selection of parking options become obsolete. With an intelligent parking solution, a new touchpoint (intelligent parking solution) is added instead. Since the customer in the consideration phase primarily only uses the intelligent parking solution including integrated navigation functionality as a touchpoint, the customer knowledge required to pass through this phase is reduced to the customer knowledge required to operate the intelligent parking solution. Whether a new touchpoint like the intelligent parking solution simplifies a customer journey or not depends on the customer. Also in the purchase phase, an intelligent parking solution adds a new touchpoint (PM11), making the parking ticket machine and the parking ticket obsolete as touchpoints (PM10). Since the customer can book a digital parking ticket via the intelligent parking solution, customer knowledge is no longer required to operate a parking ticket machine in this phase. Furthermore, when a digital parking ticket is booked, the parking information is stored centrally in the app and no longer on a paper parking ticket. This results in a reduction of the required customer knowledge, especially in the usage phase, as the customer does not need to remember the time when the parking time expires. The reduction of the required customer knowledge in this phase is based on the automation of a customer activity (PM5) in addition to the change of contact points, which reminds the customer in time that the parking time is about to expire. A reduction of required customer knowledge can also be observed in the awareness phase, which can be derived, for example, from a combination of active-passive changes (PM4) and the addition of a new touchpoint (PM11).

Another potential factor is the improvement of situational knowledge (PF7). This refers to the fact that the information provided to a customer enables a better understanding of a particular situation, which enables the customer to make better decisions or take better actions. For example, a customer's situational knowledge can be improved by adding transparency-creating (PM7), site-specific (PM8), or time-specific (PM9) information that was not previously available. For example, the situational knowledge of the customer is improved in the consideration phase by adding information that creates transparency about the parking situation at certain times in certain locations. Also in the usage phase, the situational knowledge of the customer is improved by the addition of time- and site-specific information, who can be navigated to the location of the parked vehicle in time before the expiration of the booked parking duration with the help of the intelligent parking solution. An improvement in situational knowledge can also be observed in the awareness and purchase phase.

The last potential factor documented in the parking scenario is the focusing of touchpoints (PF8). This refers to the concentration on a touchpoint in a phase that plays a central role for the customer in the execution of customer activities or sub-processes from the beginning to the end of a phase. This potential factor can be derived, for example, from combinations of the potential patterns PM7, PM8, PM9, PM10 and PM11. In the consideration phase, the focusing of a touchpoint is particularly evident. The addition of a intelligent parking solution as a new touchpoint (PM11) leads to consolidating the functionalities of original touchpoints such as navigation systems, word-of-mouth communication, social media or blogs and bulletin boards. In this case, in particular, the information functionality of the original touchpoints is consolidated in the intelligent parking solution. The intelligent parking solution experiences a high level of customer attention due to the addition of transparency-creating (PM7), site-specific (PM8) and time-specific (PM9) information, and at the same time becomes a central touchpoint of the consideration phase because original touchpoints become obsolete (PM10). Road signs and advertising banners remain in the consideration phase, but are of much less importance than the intelligent parking solution for the execution of customer activities in this phase. The focus of a touchpoint can also be observed in the purchase, awareness and usage phases, as well as to some extent in the retention phase.

6.3.3   Identified Effects Using the Example of Intelligent Parking

The derived potential factors represent further differences between the customer journeys. Effects result from different combinations of factors. Effects are changes that AI causes in certain phases of a customer journey and that occur repeatedly in different scenarios [19]. In the following, the derived potential effects of this scenario are explained. To refer to the potential effects, the numbers (see column "No." in Table 6.3) are used. The scenario over-arching effects are defined in Sect. 6.4.3.

The first potential effect resulting from the comparison of the customer journeys is a more needs-oriented accompaniment of awareness raising (PE1) in the awareness phase. This means that the customer is supported with information that makes him aware of possible parking options that are geared to his needs or that enable him to obtain a preliminary orientation about the parking situation that is geared to his needs. Improved situational knowledge (PF7) about the parking situation at the destination contributes significantly to this potential effect. The customer obtains improved situational knowledge, for example, via a focused touchpoint (PF8). The acquisition of improved situational knowledge is also associated with a phase extension (PF2).

Another potential effect in the awareness phase is a faster and more comfortable awareness raising (PE2). This potential effect can be derived, for example, from the fact that the throughput time of the awareness phase is reduced (PF3) despite a phase extension (PF2), the manual customer effort (PF5) and the required customer knowledge (PF6) are reduced. One reason for this is that the customer no longer actively searches for parking options at the destination, as he aims to use an intelligent parking solution. Nevertheless, a pre-orientation is helpful with regard to the parking search with such a parking solution to ensure that it also has an information offer at the desired destination. The time and manual customer effort for the pre-orientation is low, since the customer can fall back on a focused touchpoint (PF8) to check the information offer of a intelligent parking solution at the destination.

In the consideration phase, a more demand-oriented creation of the purchase intention (PE3) is to be mentioned as a potential effect. In the customer journey in Fig. 6.3, the customer gains improved situational knowledge (PF7) about the parking situation at the destination via the focused and central touchpoint 'Intelligent Parking Solution' (PF8). Phase extension (PF2) also plays a role in this process. Due to the improved situational knowledge, the customer is able to select an available parking space in a parking area that suits his needs best. In the conventional use case, a demand-oriented creation of the

Table 6.3 Potential effects in the parking scenario

No.	Designation of potential effects	Appearance in phases
PE1	More needs-oriented accompaniment of awareness raising	A
PE2	Faster and more comfortable awareness raising	A
PE3	More demand-oriented creation of the purchase intention	C
PE4	Faster and more comfortable creation of the purchase intention	C
PE5	More demand-oriented purchase decision	P
PE6	Faster and more comfortable purchase decision	P
PE7	Information-based extension of the product or service benefit	U
PE8	Focusing on the actual usage with fewer process-related distraction	U
PE9	Optimization of customer service	R
PE10	Strengthening of customer dialogue	R

purchase intention is not possible, since there is no transparency about the parking situation at a certain time at a certain place.

In addition, the potential effect of a faster and more comfortable creation of the purchase intention (PE4) can be observed in the consideration phase. PE4 is caused, among other things, by reducing the throughput time of the consideration phase (PF3) and reducing the manual customer effort (PF5) and the required customer knowledge (PF6). Phase extension (PF2) is also an important potential factor. In addition, the improved situational knowledge (PF7) about the parking situation contributes to the customer's ability to select a parking area with predicted available parking spaces more quickly and conveniently, and to navigate directly to it. As the customer obtains the improved situational knowledge quickly and conveniently via the intelligent parking solution, the focusing of a touchpoint (PF8) also contributes to this potential effect.

As previously explained, in the consideration phase, an intelligent parking solution allows the customer to select a parking area with predicted available parking spaces that best fits the customer's need. Thus, the customer creates more demand-oriented purchase intentions (PE3), which can result in a more demand-oriented purchase decision when the customer decides to book a digital parking ticket for a selected parking space in the purchase phase. A demand-oriented purchase decision (PE5) can be seen as a potential effect of the purchase phase, which can be derived, for example, from a combination of the potential factors PF7 and PF8. The reason for this is that the consideration and purchase phases blur into each other in this scenario, because the customer makes a kind of purchase decision already in the consideration phase by leaving the vehicle in a parking lot, although the actual purchase action does not take place until a digital parking ticket is booked in the purchase phase. The purchase decision, which takes place immediately after the selection of an available parking space, is in this respect also based on the improved situational knowledge (PF7) about the parking situation acquired via the focused touchpoint 'Intelligent Parking Solution' (PF8). It is therefore more demand-oriented than in the conventional use case.

Analogous to the potential effect PE4, a faster and more comfortable purchase decision (PE6) can be observed in the purchase phase. This potential effect can be derived, for example, from combinations of phase narrowing (PF1), reduction of throughput times (PF3) and reduction of manual customer effort (PF5) as well as the required customer knowledge (PF6). In addition, the focusing of touchpoints (PF8) supports a faster and more comfortable purchase decision, since the entire booking of the digital parking ticket takes place via the intelligent parking solution.

In the usage phase, an information-based extension of the product or service benefit (PE7) can be observed as a potential effect. This means that an intelligent parking soltion offers the customer a better parking experience in the usage phase through the targeted provision of situation-oriented information. PE7 can be derived from PF7 and PF8, among others. For example, with the reminder about the soon-to-expire parking time and the possibility to be navigated to the location of the parked vehicle, the customer receives time-specific and site-specific information from the intelligent parking solution as a central touchpoint (PF8), which improves his situational knowledge (PF7). This type of information leads to a benefit extension of the parking service.

Another potential effect in the usage phase is the focus on the actual usage with less process-related distraction (PE8), by which is meant that the customer can concentrate more on using the parking service when passing through the usage phase. This potential effect can be derived, for example, from combinations of the potential factors PF1, PF5, PF6 and PF8. For example, the phase narrowing (PF1) as well as the focus on the intelligent parking solution as a central touchpoint (PF8) have the effect that the customer benefits faster from the paid parking time ("park and leave vehicle"). The reduction of manual customer effort (PF5) and required customer knowledge (PF6) also result in the customer being exposed to less distraction while parking. For example, the customer no longer needs to regularly check the remaining parking time, which allows him to focus more on the activities during parking (e.g. keeping an appointment).

A potential effect in the retention phase is the optimization of customer service (PE9). This means that the forecasts of the real-time parking situation can be directly improved by the intelligent parking solution based on customer ratings. For example, PE9 can be derived from combinations of the potential factors PF2, PF4, PF7 and PF8. For example, phase extension (PF2), extension of throughput times (PF4), and focusing of touchpoints (PF8) all lead to customer evaluations of the parking forecast with available parking spaces. Based on customer ratings, AI algorithms learn to better assess the real-time parking situation and provide better information to the customer when using the intelligent parking solution again (PF7). By allowing customer reviews to help improve the intelligent parking solution's product features, customers can expect a better parking experience when they use it again. The satisfaction level of a dissatisfied customer can be increased if the customer has a better experience with the intelligent parking solution when using it again.

The last identified potential effect in this scenario is the strengthening of customer dialogue (PE10) in the retention phase. This means that the interaction with the customer is maintained after the use of the parking service. Specifically, the customer is given the opportunity with the intelligent parking solution to evaluate the predicted availability of parking spaces in a parking area. This potential effect can be derived, for example, from combinations of phase extension (PF2) and extension of throughput times (PF4). Furthermore, the focus of the central touchpoint 'Intelligent Parking Solution' (PF8) contributes to the derivation of this potential effect.

With the identification of eleven potential patterns as well as the derivation and identification of eight potential factors and ten potential effects, the main differences between the customer journeys are documented in Figs. 6.2 and 6.3. In order to find out whether the potential patterns, factors and effects identified and derived in the parking scenario also occur across scenarios, customer journey comparisons were carried out for six other scenarios following the demonstrated procedure. In Sect. 6.4, the results of the customer journey comparisons of all seven scenarios are generalised to determine which potential patterns, factors and effects are scenario overarching.

6.4   Generalisation and Discussion of the Results

In Sect. 6.3, customer journeys for the parking scenario were compared with each other and differences between the customer journeys were identified. Analogous to this procedure, customer journeys from six other scenarios were modeled, compared with each other and differences documented. With the documented differences, potential patterns have been identified for a total of seven scenarios, and potential factors and effects have been derived. In this section, the objective is to generalize the documented differences across all scenarios and identify intersections. Specifically, for each potential pattern, factor and effect, it is evaluated in which scenarios and in which phases the potential pattern , factor or effect occurs. It is then determined which potential patterns, factors and effects are actually patterns, factors, and effects based on the appropriate definitions. In the following, all potential patterns in Sect. 6.4.1, potential factors in Sect. 6.4.2 and potential effects in Sect. 6.4.3 are generalised.

6.4.1   Patterns Are the Most Granular Changes Between Customer Journeys of an AI-Based and Traditional Use Case

Patterns are changes that AI causes in customer activities or customer sub-processes or touchpoints and that occur repeatedly in different phases of a customer journey and in numerous scenarios [19]. A total of 13 potential patterns (PM1 to PM13) were identified from scenarios S1 to S7, which are generalized in Table 6.4. For all potential patterns, it is summarized in which scenarios and in which phases of a customer journey they were identified.

In the following, it must be determined which potential patterns are actually patterns according to the above-mentioned definition. Table 6.4 shows that the potential patterns PM1 to PM12 were each identified in several scenarios and occur repeatedly in at least two phases of a customer journey. Only the potential pattern PM13 solely occurs in one scenario and the consideration phase. Accordingly, the potential patterns PM1 to PM12 are indeed patterns. To refer to these patterns in the following, the numbers M1 to M12 are used analogously to the previous numbers PM1 to PM12. Based on the customer journey comparisons performed, PM13 cannot be determined as a pattern as it has not been identified in multiple scenarios and only occurs in one phase. In order to investigate whether PM13 is actually a pattern, customer journey comparisons of further scenarios must be carried out.

6.4.2   Factors Are Derived from Pattern Combinations

Factors are derived from different combinations of patterns and are changes that AI causes in different phases of a customer journey and that occur repeatedly in numerous scenarios [19]. From the patterns M1 to M12 described in Sect. 6.4.1, a total of ten potential factors were derived and identified in scenarios S1 to S7, which are generalized in Table 6.5. For

Table 6.4 Generalization of the identified potential patterns

No.	Designation of potential samples	Performance in scenarios	Appearance in phases
PM1	Elimination of existing customer activities/sub-processes	S1, S2, S3, S4, S5, S6, S7	C, P, U
PM2	Postponement of existing customer activities/sub-processes	S1, S3	C, P, U
PM3	Addition of new customer activities/sub-processes	S1, S2, S3, S4, S5, S6, S7	A, C, P, U, R
PM4	Active-passive change for existing customer activities/sub-processes	S1, S2, S3, S4, S5, S6	A, C, P, U, R
PM5	Automation of customer activities/sub-processes	S1, S2, S3	C, P, U, R
PM6	Parallelization of customer activities/sub-processes	S1, S2	C, P, U
PM7	Addition of transparency-creating information	S1, S3, S4, S5	A, C, P
PM8	Addition of site-specific information	S1, S3, S5	A, C, P, U
PM9	Addition of time-specific information	S1, S4	A, C, P, U
PM10	Status change from original to obsolete touchpoints	S1, S2, S3, S4, S5, S6, S7	A, C, P, U, R
PM11	Addition of new touchpoints	S1, S2, S3, S4, S5, S6	A, C, P, U, R
PM12	Addition of customer-oriented information	S4, S5	C, P
PM13	Elimination of existing intermediate events	S5	C

all potential factors, it is summarized in which scenarios and in which phases of a customer journey they were derived and identified.

Analogous to the previous Sect. 6.4.1, it is now necessary to determine for all potential factors PF1 to PF10 which potential factors are actually factors according to the above-mentioned definition. As shown in Table 6.5, all ten potential factors occur repeatedly in several scenarios and in different phases of a customer journey. Therefore, it can be concluded that all potential factors PF1 to PF10 are actually factors. Analogous to the numbers of the potential factors PF1 to PF10, the numbers F1 to F10 are used below for these factors.

6.4.3 Effects Are Derived from Factor Combinations

Effects are derived from different combinations of factors and are changes that AI causes in certain phases of a customer journey and that occur repeatedly in different scenarios [19]. In the scenarios S1 to S7, a total of 19 potential effects were derived and identified from the factors F1 to F10 described in Sect. 6.4.2, which are generalized in Table 6.6. For all potential effects, it is summarized in which scenarios and in which phases of a customer journey they were derived and identified.

Table 6.5 Generalization of the derived and identified potential factors

No.	Designation of potential samples	Performance in scenarios	Appearance in phases
PF1	Phase narrowing	S1, S2, S3, S6, S7	C, P, U
PF2	Phase extension	S1, S3, S4, S6	A, C, P, R
PF3	Reduction of throughput times	S1, S2, S3, S4, S5, S6, S7	A, C, P, U, R
PF4	Extension of throughput times	S1, S4, S6	P, R
PF5	Reduction of manual customer effort	S1, S2, S3, S4, S5, S6, S7	A, C, P, U, R
PF6	Reduction of required customer knowledge	S1, S2, S4, S5, S7	A, C, P, U, R
PF7	Improvement of situational knowledge	S1, S3	A, C, P, U
PF8	Focusing of touchpoints	S1, S2, S3, S4, S5, S6, S7	A, C, P, U, R
PF9	Online-offline channel linking	S3, S5, S6	C, P
PF10	Personalization and improvement of situational knowledge	S4, S5	C, P

As in the previous Sect. 6.4.2, it is now necessary to determine for all potential effects PE1 to PE19 which potential effects are actually effects according to the above definition. As can be seen in Table 6.6, the potential effects PE1 to PE6, as well as PE8, PE10, PE11 and PE13 occur repeatedly in different scenarios. Consequently, it can be concluded that these potential effects are indeed effects. Therefore, to refer to these effects, the numbers E1 to E6, as well as E8, E10, E11 and E13 are used. Based on the customer journey comparisons performed, potential effects PE7, PE9, PE12, and PE14 through PE19 are not effects because they do not occur in multiple scenarios. Additional customer journey comparisons of further scenarios are required to investigate whether these potential effects are indeed effects. In summary, a total of ten effects and nine potential effects were derived and identified that represent (potential) characteristic, qualitative process-oriented influences of AI on customer journeys and are structured and documented in the framework proposed by Schneider et al. [19] (Fig. 6.1).

Table 6.6 Generalization of the derived and identified potential effects

No.	Designation of potential effects	Performance in scenarios	Appearance in phases
PE1	More needs-oriented accompaniment of awareness raising	S1, S3	A
PE2	Faster and more convenient awareness raising	S1, S3	A
PE3	More demand-oriented creation of the purchase intention	S1, S3, S4, S5	C
PE4	Faster and more comfortable creation of the purchase intention	S1, S2, S3, S4, S5, S6, S7	C
PE5	More demand-oriented purchase decision	S1, S3, S4, S5	P
PE6	Faster and more comfortable purchase decision	S1, S2, S3, S5, S6	P
PE7	Information-based extension of the product or service benefit	S1	U
PE8	Focus on actual useage with fewer process-related distractions	S1, S2	U
PE9	Optimization of customer service	S1	R
PE10	Strengthening of customer dialogue	S1, S2	R
PE11	Increased discretion	S3, S5	C, P
PE12	Permanent service availability	S3	C, P
PE13	Customer trust	S3, S5, S7	R
PE14	Satisfaction of other needs	S4	P
PE15	Versatile language support	S5	C, P
PE16	Individual, digital consulting in the physical world	S5	C, P
PE17	Loyal customer relationship	S6	R
PE18	Access to services independent of knowledge and ownership	S7	C
PE19	Increased security	S7	C

6.5   Conclusion and Outlook

New applications of AI are increasingly emerging in the consumer market. Increasingly, devices and services that communicate autonomously over the internet are also entering the market. This can enhance these devices and services with novel AI-based services. Such services can influence the way customers make commercial decisions.

Against this background, a growing interest in the design of customer journeys in connection with AI has developed in the CEM environment. Research on the influence of AI on customer journeys, on the other hand, is still in its early stages. Especially on process level, the combined consideration of AI and customer journeys represents a gap in research. To address this novel research area, Schneider et al. propose a framework [19] that provides an initial overview of the largely unexplored effects of AI on commercial interactions. Building on [19], this chapter demonstrated how these effects are derived from so-called patterns and factors identified in numerous customer journey comparisons.

The exemplary representation of the parking scenario illustrates the procedure for deriving the patterns, factors and effects. The necessary level of detail is clarified. From a scientific point of view, the results presented contribute to the development of this novel field of research. The approach can also serve as a guide for further research building on this chapter and on [19], which is also concerned with an investigation of the influences of AI on commercial interactions. Thus, this chapter provides a good starting point for further research activities.

Future research can, for example, address the empirical study of customer journeys to optimize the quality of the approach to inferring the effects of AI on commercial interactions. Empirically collected data can be used to limit interpretive leeway in the combinations of patterns used to derive factors and factors used to derive effects. Analyses of further scenarios following the demonstrated procedure may also improve the database, potentially further reducing interpretive freedoms in the derivation of effects.

In summary, it can be said that AI can significantly influence customer journeys. The important thing here is to work out the changes between AI-based and conventional processes and to achieve corresponding improvements perceived by the customer. In other words, it is not about pure automation of manual customer processes, but about a holistic redesign of the customer experience. The patterns, factors and effects presented here can be used in practice as a basis for understanding this potential of AI and implementing it in the design of one's own customer journeys.

References

1. Scholz C (2017) Zukünftige Herausforderungen in der Personalentwicklung und die sich daraus ergebenden neuen Rollen. In: Covarrubias Venegas B, Thill K, Domnanovich J (Hrsg) Personalmanagement: internationale Perspektiven und Implikationen für die Praxis. Springer Gabler, Wiesbaden, S. 411–432
2. Wolfgang R (2017) Nachhaltigkeit. Nachhaltigkeit im Projektmanagement. Springer Gabler, Wiesbaden, S. 7–33
3. Rusnjak A, Schallmo DRA (2018) Gestaltung und Digitalisierung von Kundenerlebnissen im Zeitalter des Kunden. In: Rusnjak A, Schallmo DRA (Hrsg) Customer Experience im Zeitalter des Kunden: Best Practices, Lessons Learned und Forschungsergebnisse. Springer Gabler, Wiesbaden, S. 1–40
4. Halvorsrud R, Kvale K, Følstad A (2016) Improving service quality through customer journey analysis. J Serv Theory Pract 26(6):840–867
5. Lemon KN, Verhoef PC (2016) Understanding customer experience throughout the customer journey. J Mark 80(6):69–96
6. Hauk J, Czarnecki C, Dietze C (2018) Prozessorientierte Messung der Customer Experience am Beispiel der Telekommunikationsindustrie. In: Rusnjak A, Schallmo D (Hrsg) Customer Experience im Zeitalter des Kunden. Springer Gabler, Wiesbaden, S. 195–216
7. Bhattacharya A, Srivastava M, Verma S (2018) Customer experience in online shopping: a structural modeling approach. J Glob Mark 32(9):1–14
8. Khan N, Akram M U, Shah A, Khan S A (2018) Calculating customer experience management index for telecommunication service using genetic algorithm based weighted attributes. In: 2018 IEEE international conference on innovative research and development (ICIRD), Bangkok, S. 1–8

9. Bruhn M, Hadwich K (2012) Customer experience—Eine Einführung in die theoretischen und praktischen Problemstellungen. In: Bruhn M, Hadwich K (Hrsg) Customer experience: forum Dienstleistungsmanagement. Springer Gabler, Wiesbaden, S. 3–36

10. Palmer A (2010) Customer experience management: a critical review of an emerging idea. J Serv Mark Emerald Bingley 24(3):196–208

11. Zinkann R, Mahadevan J (2018) Zukünftige Customer Journeys und deren Implikationen für die Unternehmenspraxis. In: Bruhn M, Kirchgeorg M (Hrsg) Marketing Weiterdenken: Zukunftspfade für eine marktorientierte Unternehmensführung. Springer Gabler, Wiesbaden, S. 157–169

12. Holmlid S, Evenson S (2008) Bringing service design to service sciences, management and engineering. In: Hefley B, Murphy W (Hrsg) Service science, management and engineering education for the 21st century: research and innovations in the service economy. Springer, Boston, S. 341–345

13. Plottek K, Herold C (2018) Micro Moments als entscheidender Moment im Rahmen einer zunehmend fragmentierten Customer Journey. In: Rusnjak A, Schallmo DRA (Hrsg) Customer Experience im Zeitalter des Kunden: best practices, lessons learned und Forschungsergebnisse. Springer Gabler, Wiesbaden, S. 144–176

14. Nölle N, Wisselink F (2018) Pushing the right buttons: how the internet of things simplifies the customer journey. In: Krüssel P (Hrsg) Future Telco: successful positioning of network operators in the digital age. Springer, Cham, S. 319–328

15. Schneider D, Wisselink F, Czarnecki C (2018) Nutzen und Rahmenbedingungen informationsgetriebener Geschäftsmodelle des Internets der Dinge. In: Barton T, Müller C, Seel C (Hrsg) Digitalisierung in Unternehmen: Von den theoretischen Ansätzen zur praktischen Umsetzung. Springer Vieweg, Wiesbaden, S. 67–85

16. Mackenzie M (2016) LPWA networks for IoT: worldwide trends and forecasts 2015–2025. Analysys Mason Limited, London

17. Bitkom DFKI (2017) Künstliche Intelligenz: Wirtschaftliche Bedeutung, gesellschaftliche Herausforderungen, menschliche Verantwortung. Bitkom e. V. & DFKI, Berlin

18. Lenzen M (2002) Natürliche und künstliche Intelligenz: Einführung in die Kognitionswissenschaft. Campus, Frankfurt a. M.

19. Schneider D, Wisselink F, Nölle N, Czarnecki C (2020) Influence of artificial intelligence on commercial interactions in the consumer market. In: Bruhn M, Hadwich K (Hrsg) Automatisierung und Personalisierung von Dienstleistungen: Forum Dienstleistungsmanagement. Springer Gabler, Wiesbaden, S. 183–205

20. Peters C, Zaki M (2018) Modular service structures for the successful design of flexible customer journeys for AI services and business models—orchestration and interplay of services. In: Working paper series of the Cambridge Service Alliance. Cambridge Service Alliance, University of Cambridge, Cambridge, S. 1–9

21. eParkomat (2018) We predict real-time parking situation. https://eparkomat.com/. Accessed 26 Mar 2020

22. Statista (2017) Durchschnittliche Suchzeit für Parkplätze in deutschen Großstädten nach Art des Parkens. https://de.statista.com/statistik/daten/studie/732264/umfrage/aufgewendete-zeit-fuer-die-parkplatzsuche-in-deutschen-grossstaedten/. Accessed 29 Feb 2020

Dominik Schneider , M.Sc is Key Account Manager at Deutsche Telekom. He helps large enterprise clients to unlock value through cutting-edge communication technologies and digitalization solutions. With a Master's degree in Business Information Systems and over ten years of professional experience encompassing sales, strategy consulting and IT management, he brings expertise in mobile communications, IoT, device lifecycle strategies and the economic implications of AI and Big Data.

Dr. Ir. Frank Wisselink , B.Sc. (Hon) is the Executive Product/Project Manager for AI of Deutsche Telekom T-Systems. His expertise covers the complete spectrum for the digitization of Europe starting with communication technologies like Narrowband IoT to sustainable value creation with AI. He is an active member in several national and international working groups and delegations at BITKOM, DIN, CEN-CENELEC & ISO for the standardization of AI within the framework of the EU AI ACT.

Nikolai Nölle is the Product Manager of "Frag Magenta", Deutsche Telekom's award-winning digital sales & service assistant in Germany. Best in class service experience and simplify customers' lives is his prime focus. He leads the product vision, innovative use cases, customer journeys, interactions or dialogs for superior customer experience. He achieves this through the consistent application of customer experience management methods and the latest AI technologies to customers' favor.

Prof. Dr.-Ing. Christian Czarnecki is Professor of Information Systems at FH Aachen University of Applied Sciences. After studying business informatics at the University of Münster, he gained more than 10 years practical experience in various consultancies and managed a large number of transformation projects in Europe, Africa and the Middle East. He received his doctorate in engineering from the University of Magdeburg in 2013. His main research interests are Digital Transformation, Process Management, Robotic Process Automation, Internet of Things and Enterprise Architectures. His work has been published in leading academic journals, at international conferences and in various books.

Artificial Intelligence (AI) for Platform Business Models

7

Gabriele Roth-Dietrich

Abstract

The further development into a platform and the use of AI technologies to support the platform business model offers companies the opportunity to counter the disruptive changes of the digital transformation. Platforms appear as business model patterns with different success rates in diverse business model clusters and also differ in terms of the way they use data. The design aspects of a platform, from launch to governance and controlling of established platforms, imply data use of varying intensity, for example for process control, decision preparation or automation. This results in different AI deployment options with varying degrees of impact on platform success depending on their influence on competitiveness and depending on the frequency of the supported processes, as well as several options for action for companies that want to locate themselves in the platform ecosystem.

7.1 Platform Economy

7.1.1 Platforms

A platform is a business model of an at least **two-sided market** that enables value-creating interactions between consumers, producers and possibly other independent user groups. Platform companies create value for all participants by organising the **coming together of users** and the **exchange of products, services or social currency,** i.e. intangible,

G. Roth-Dietrich (✉)
Hochschule Mannheim, Mannheim, Deutschland
e-mail: g.roth-dietrich@hs-mannheim.de

subjective values. Digital platform companies use digital technologies for the smooth platform access of all participant groups and the matching of suitable interaction partners even
with a large user crowd. Utility and platform value typically increase with the number of
user groups or individual group members. The special feature of platforms is the distribution of **value creation** away from the company to the **user community of** the platform, so
that a platform company does not need to own or control the **resources** to achieve value
creation (cf. [1, p. 26ff.]).

In a **platform ecosystem,** as shown in Fig. 7.1, actors interact with different roles. The
platform owner controls the intellectual property and defines the rules of the game on the
platform, such as who may participate and which interactions are possible. The **platform operator** (provider) is the intermediary of user interactions and the primary point of contact between
the platform users and the platform. **Producers** provide offers on the platform that **consumers**
demand. Both exchange values and data via the platform and provide feedback (cf. [2], [
3, p. 2]).

Several factors contribute to the **disruptive effect of platforms** that drive established
companies out of the market. In the traditional pipeline business model, value is created
step by step from supplier to manufacturer to customer in a linear, self-controlled value
chain (pipeline). Platform companies no longer use the Internet merely as a distribution
channel, but as a generating infrastructure and coordination mechanism. In addition, there
is the rapid convergence of the real and digital worlds into an Internet of Things (IoT),
where physical objects connect to and are controlled by the Internet. Platform companies
are taking advantage of this and creating entirely new business models. As platforms rely
on external ecosystems to create value, they have key advantages in the growth phase, such
as cheaper marginal costs when expanding offerings. Network effects further accelerate
growth so that platforms can achieve significantly higher value creation than pipeline companies (cf. [1, p. 69ff.]).

7.1.2   Network Effects

While companies in the industrial age achieved great growth through supply-side economies of scale, platforms benefit from demand-side economies of scale, the so-called **network effects,** which make larger networks more attractive for participants (**Metcalfe's**

Fig. 7.1 Players in the
platform ecosystem

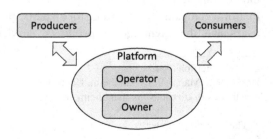

law). One-sided network effects are triggered by users on one side of the market and affect users on the same side of the market, while two-sided effects affect users on the other side of the market. An example of **one-sided positive effects** is the benefit of the telephone network, which grows disproportionately with the number of connections. **Positive two-sided effects** tap into sales platforms, for example, where an increase in the number of buyers attracts more sellers because they suspect interesting sales opportunities due to high demand, while conversely a large number of sales offers attracts more users to the platform because the greater choice increases the probability of suitable offers. These cross-page effects act like **positive feedback loops** (cf. [1, p. 27ff.]).

Negative network effects can occur if the platform has already grown considerably and it becomes confusing for the community to find suitable offers. If the platform participants receive more and more notifications about potential offers, e.g. in the form of advertising or suggestions, this overload may have a deterrent effect. Sometimes initially positive effects with unequal growth of the two market sides later turn into negative ones, the attractiveness of the platform suffers, and **negative feedback loops** slow down further growth because dissatisfied users turn their backs on the platform (cf. [1, p. 37ff.]).

7.1.3 Business Model Patterns for Platform Companies

A business model shows in abstract terms how a company can generate added value for customers and secure a return (see [4]). It addresses the target customers and the value proposition to them, the value chain for the provision of services and the revenue mechanism (cf. [5, p. 6f.]). The St. Gallen Business Model Navigator outlines a multi-stage iterative procedure to develop business model innovations through environment analysis, idea generation and business model design. It recommends creative imitation of other successful companies, recombination of **business model patterns** found there and their transfer, adaptation and design in one's own organization (cf. [5, p. 21f.]). The 55 **business model patterns function as neutral** design aids, since they are defined in a generally valid way independent of industries and company sizes and can therefore be applied in every company (cf. [6]). Gassmann et al. list platforms as the business model pattern **Two-Sided Market**, which brings together two distinguishable user groups on the platform of a third party, but also mention multi-sided markets with three or more user groups, such as search engines, which connect searchers, advertisers and website operators. The central challenge for platform operators is to manage the user groups in order to maximize the positive network effects (cf. [5, 7]).

In a quantitative empirical study, Böhm et al. code the 55 business model patterns used or not used by the companies considered as the business model DNA of the organization. They extract successful pattern combinations that have been tested in practice and bundle them into twelve **business model clusters** that predict a probability-based statement on the goodness, probability of success and chances of survival of the business models. Each business model cluster combines several business model patterns (cf. [8, p. 1009ff.]).

In the course of clustering, Böhm et al. first summarize the application of one or more of the business model patterns Two-Sided Market, Orchestrator or Long Tail as the **business model pattern Platform** (cf. [8, p. 1011f.]). To the approach of the multi-sided markets, they thus add the **orchestration task of** the platform operator, who coordinates value-adding activities of diverse companies like a conductor in order to be able to offer users a cleverly combined and aggregated range of services. The company concentrates on its core competencies and outsources all other process steps to specialized service providers. The high transaction costs are offset by potential savings through specialization advantages (cf. [5, p. 252ff.]). The **Long Tail** pattern refers to the offering and selling of niche products and thus distances itself from the classic 80–20 rule, according to which companies generate 80% of their sales from 20% blockbuster products. Users benefit from a more colorful range of products and have the chance to satisfy more unusual demand (cf. [5, p. 224ff.]). According to Böhm et al. the aggregated pattern Platform or the basic pattern Two-Sided Markets can be found in six of the twelve business model clusters. This demonstrates the high importance of the platform idea. In their investigation of the **success of the clusters,** however, clear differences emerge in the survival rate and in the proportion of fast or slow growing companies in the clusters (cf. [8, p. 1013]). Table 7.1 shows the results of Böhm et al. for the platform-related clusters and adds a new score value for the cluster success, which, due to the great importance of platform growth, weights the rate for high growth twice and the survival rate once.

The worst score was achieved by the **Hidden Revenue Markets** cluster, which combines Two-Sided Markets, Affiliation, Long Tail and the Hidden Revenue pattern. Affiliate marketing relies on the financing of free Internet offers by third parties paying fees for customer acquisition on the platform, comparable to price comparison portals or brokerage commissions for independent insurance agents (cf. [5, p. 100]). Hidden Revenue generates revenues not by selling the services provided, but by commercializing an advertising space. Companies in this cluster serve both traditional and advertising customers and can use advertising revenues to cross-subsidize products that they offer at a discount or even for free (cf. [5, p. 190]). Overall, the cluster does not appear to be very promising for platforms, possibly because platform users are sceptical about an excess of paid advertising offers.

Table 7.1 Success of platform-related business model clusters

Business model cluster	High growth rate [%]	Survival rate [%]	Score
Hidden Revenue Markets	25	75	50
Freemium Platforms	50	100	67
Mass Customizing Orchestrators	50	83	67
Affiliate Markets	52	78	68
Innovative Platforms	67	100	**78**
Crowdsourcing Platforms	73	91	**82**

Freemium Platforms, Mass Customizing Orchestrators and Affiliate Markets perform better. The **Freemium Platforms** cluster includes companies that offer priced services in addition to free basic services. However, the low growth rates suggest that only a few platform customers are willing to pay for additional services (see [8, p. 1012f.]).

With **Mass Customizing Orchestrators,** companies combine the Orchestrator and Two-Sided Markets patterns with the Layer Player pattern and with Mass Customization. A layer player acts as a layer specialist, takes over only a few activities of the value chain, interacts mostly with orchestrators and can thus multiply its know-how (cf. [5, p. 203]). Mass customization resolves the contradictions between mass production and customer individualization. Modularized product hierarchies make it possible to assemble many differentiated end products tailored to individual customer needs from standardized modules with a high degree of efficiency (cf. [5, p. 233]). Mass customizing orchestrators also show only low growth rates, possibly because the self-responsible share of value creation scales too slowly with growth.

Affiliate Markets complement the platform model with the Affiliation and Aikido business model patterns. Aikido converts the opponent's strengths into weaknesses. In the Aikido business model, companies offer services that are fundamentally different from those of the competition, thus bypassing direct comparison, can grow comfortably in an almost competition-free zone, and later surprise the competition, as their advantages then no longer stand up to the new offer (cf. [5, p. 105]). Both survival and growth rates of affiliate marketers rank in the middle. This shows that in practice it can be difficult firstly to find suitable niches and secondly to conquer larger market shares from these later on.

The highest score values are achieved by Innovative and Crowdsourcing Platforms. **Innovative Platforms** combine the patterns Aikido, Two-Sided Market, Orchestrator and Revenue Sharing. In the latter case, companies share their revenue with stakeholders who have played a significant role in generating the revenue, for example through customer recommendations or by taking on value-adding activities in the service creation process (cf. [5, p. 282]). **Crowdsourcing Platforms** extend platforms and Aikido with crowdsourcing and customer loyalty. While crowdsourcing means outsourcing part of the activities in the course of tendering to external actors in order to expand one's own knowledge horizon and to arrive at cost-effective and efficient solutions to problems (cf. [5, p. 132]), customer loyalty describes incentive systems for customer loyalty, such as card-based bonus programs that reward sales through points that customers can convert into rewards in kind or credits. Psychological effects are designed to induce users to make purchases from rewards program providers. Incentivization produces lock-in effects (cf. [5, p. 137f.]). Since crowdsourcing requires an active community, the patterns complement each other perfectly. In addition, customers are increasingly no longer acting as mere consumers, but are seeking dialogue with companies and are willing to contribute time and energy to crowdsourcing campaigns. The Crowdsourcing Platform cluster even shows the highest growth rates of all platform-related clusters. The superior performance of the two clusters shows the high importance of binding participants to the platform, opening the platform to contributions from external third parties, and fairly distributing the added value generated by the platform. Skillful design of the platform can therefore decisively increase its chances of success.

7.2 Artificial Intelligence (AI) for Platform Companies

7.2.1 AI Sub-sectors

Artificial Intelligence (AI) is a branch of computer science that explores mechanisms of intelligent human behavior by simulation using artificial artifacts such as computer programs on a computing machine. This general definition has its limitations, as the term intelligent human behavior itself has not yet been adequately defined (cf. [9]). AI is concerned with intelligent problem-solving behavior and the creation of intelligent computer systems that enable a computer to solve tasks that would require human intelligence to solve (cf. [10]). **Strong AI** would have the same intellectual abilities as humans, could support them in dealing with difficult tasks and cooperate with them on an equal footing, would act flexibly on its own initiative, but according to expert opinion does not exist as of today (cf. [11, p. 14]).

Weak AI focuses on coping with concrete application problems in individual areas and has the integral ability to learn and optimize itself. Weak AI applications simulate intelligent behavior using methods from mathematics and are able to deal with uncertainty and probability data (cf. [12]). Weak AI learns through **machine learning (ML),** i.e., methods that learn without being explicitly programmed to do so, and that optimize the problem-solving path itself. In the simplest case, **supervised learning,** the system is given data where both inputs and results are known. The ML system adjusts the algorithm until the input and output match, using it to solve, for example, regression and classification problems. **Unsupervised learning** has no desired output available and looks for patterns in unstructured data, such as in clustering tasks. **Reinforcement learning** steers the algorithm in the desired direction with pluses and minuses. In **Deep Learning,** neural networks mimic the structures of the human brain (cf. [11, p. 20ff., 13, 14]). In the following, the investigation focuses on machine learning as a subarea of AI and therein in particular on application possibilities for supervised and unsupervised learning in platforms.

7.2.2 Types of AI and Intensity of Data Use

Companies use different types of AI for different purposes. In a broad definition, AI subsumes not only systems that can sense and learn from their environment, but also rule-based systems, for example to automate manual or cognitive routine and other tasks without creating novel solution paths **(Automation). Assisted Intelligence** AI systems, which are also rigid, help humans with task completion or decision making. Unlike these two types of AI in more hard-wired systems, adaptive AI systems improve decision-making (**Augmented Intelligence**) by permanently learning from users and the environment. Finally, **Autonomous Intelligence** automates decision-making processes without human involvement and can independently adapt to different situations (cf. [15, p. 2]).

Depending on the type of AI, business models with different **intensities of data use** open up for a company (cf. [16]):

- **Low Data Business Models** hardly use digital technologies and collect data only to a small extent. The companies provide products or services in the traditional way and only use IT tools to control their processes, e.g. in ERP or CRM systems.
- **Data-Enhanced Business Models** combine service provision with digital technologies to optimize products or services. For example, companies can make a rental offer usable via an app.
- In **Pure Data-Driven Business Models,** data is the key resource for offering digital products or services. Data processing, aggregation and analysis processes generate new value for customers. This is where the USPs (Unique Selling Proposition) of platform business models can be found.
- It is possible that the Pure Data-Driven Business Model will evolve into, or be replaced by, a **Deep-Learning Business Model,** whereby powerful AI uses data to develop a self-learning intelligent business model without human intervention.

Figure 7.2 shows the types of AI required for business models with different data usage intensities. Low Data or Data-Enhanced Business Models already benefit from assisted intelligence and improve their chances of success in the market through process acceleration or automation. Optimizing day-to-day business processes is also imperative in platforms as they manage the interactions of a large number of market participants.

	with human interaction	without human interaction
rigid systems	**Assisted Intelligence** Low Data Business Models Data-Enhanced Business Models Pure Data-Driven Business Models	**Automation** Low Data Business Models Data-Enhanced Business Models Pure Data-Driven Business Models
adaptive systems	**Augmented Intelligence** Pure Data-Driven Business Models	**Autonomous Intelligence** Pure Data-Driven Business Models Deep-Learning Business Models

Fig. 7.2 AI types for business models of different data usage intensity

Furthermore, Augmented Intelligence promotes the process quality of platform interactions. Ultimately, platform companies should increasingly evolve towards Autonomous Intelligence to learn more and more based on their experiences, like Deep-Learning Business Models, and evolve their business model.

7.2.3   AI Deployment Opportunities in Platform Companies

AI influences platform providers on the one hand through **productivity increases** due to **internal process improvements or redesigns** that either relieve employees of routine tasks and give them room for more demanding and higher-value activities or that replace human labor entirely (cf. [17, p. 47]). To **business strategy,** AI supports decision-making on offerings, pricing, and marketing strategies, e.g., through simulations on market conditions, demand forecasts, and pricing. AI contributes to **research, development and innovation** by accelerating the discovery of new trends. AI enriches **marketing, sales and service** by increasing customer loyalty and conversion rates on the basis of reduced information asymmetries between supplier and customer and the precise tailoring of communication. Examples include personalized recommendations of products and services, chatbots as customer service agents, analysis of emotions in call centers, and monitoring of sales practices. Support functions such as accounting, IT and risk management benefit from AI through cost savings and risk reduction through better planning and forecasting, for example in the calculation of probabilities for payment defaults by defaulting platform users (see [18, p. 11ff.]). Since the classic value-adding areas such as **procurement, production, supply chain and logistics** are usually the responsibility of the platform producers and not the provider, the providers benefit here, for example, through on-demand production based on AI demand estimates or through autonomous reordering of raw materials depending on the level of platform activity.

A second field of action of AI on platforms is the intra- or cross-industry **improvement or new development of products and services** for customers (cf. [17, p. 47]). **Product personalization requires** that companies have precise insight into the wishes and preferences of their customers. AI enables platform providers to access and correlate data points across the entire customer journey and draw conclusions about consumption drivers. In this sense, AI bridges the gap between Big Data data collection and mass customization. ML algorithms enable risk-free test-and-learn approaches and insights into customized recommendations for customers or individualized product design. In addition, AI increases product value for users, for example by expanding functionality or in the form of improved **product quality.** By evaluating the collected preferences of many users, AI systems can derive recommendations for new product features or services or even control the product development and production process. For buyers, AI tools save time when searching for suitable offers and overcome breaks in the purchasing process (cf. [18, p. 11ff.]).

7.3   Importance of AI for Design Aspects of Platform Companies

The following study analyzes the importance of AI for various aspects of the design of a platform company and distinguishes, on the one hand, the relevance for platform success or the **competitive advantages** that AI use of the platform opens up through differentiation from competitors and, on the other hand, the **process frequency** for AI application (cf. Fig. 7.3). Application scenarios that are already taken for granted and widely used, such as rule-based systems for Assisted and Automation Intelligence, do not represent a differentiating feature and should be considered a commodity if the process frequency is high and a nice-to-have if it occurs less frequently. AI deployments, which are currently useful for differentiation from competitors, are to be classified as an essential key resource for platforms, regardless of the frequency with which they are used. The greatest competitive advantage is offered by use cases that are still hardly used in platforms. With a high degree of use, these are the strategic potentials for the platform; with a low degree of use, they at least create desirable scenarios for the establishment of a niche offering.

7.3.1   Hybridisation of Business Models

Platforms can build a different **relationship of physical assets and digital platform eco-system. Asset-heavy platforms** primarily operate like a classic pipeline company, but are

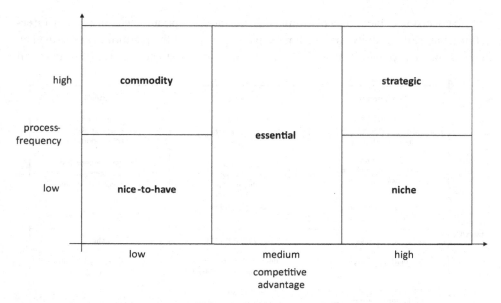

Fig. 7.3 Subdivision of AI application scenarios according to process frequency and competitive advantage

starting to test their own platforms. **Asset-light companies** are dominated by the platform approach and have little or no physical assets (see [19, p. 42]). **Mixed forms** combine pipeline activities with platform activities in the sense of complementing each other. Overall, there seems to be a trend towards the mixed area, into which many companies are slowly moving from both extremes.

As pipeline companies experiment with platform approaches, this increases the importance of AI as the **amount of data to be analyzed** increases due to the resulting more data-heavy business model. In addition, this **hybridization of business models** opens up new opportunities for AI use, for example to beneficially transfer customer or product-related data from the pipeline model to the platform business. Conversely, from the **cross-business model data use** of platform data in the company, they gain new insights for the further development of their traditional fields of activity. The potential for **competitive advantages** from the joint use of pipeline data is high, as this opportunity is denied to pure platform companies. However, it remains open to what extent the transfer of data makes sense, as customer and usage profiles in the traditional and platform businesses may diverge or as the business models address incomparable customer groups. Nevertheless, in **niche areas,** sharing the data could achieve competitive insights (see Fig. 7.4).

7.3.2 Design of Suitable Platform Processes

The first decision in building a platform concerns the appropriate selection of a **key interaction,** that is, the activity for which most participants visit the platform. During the key interaction, the provider provides values that the customer consumes. Platform

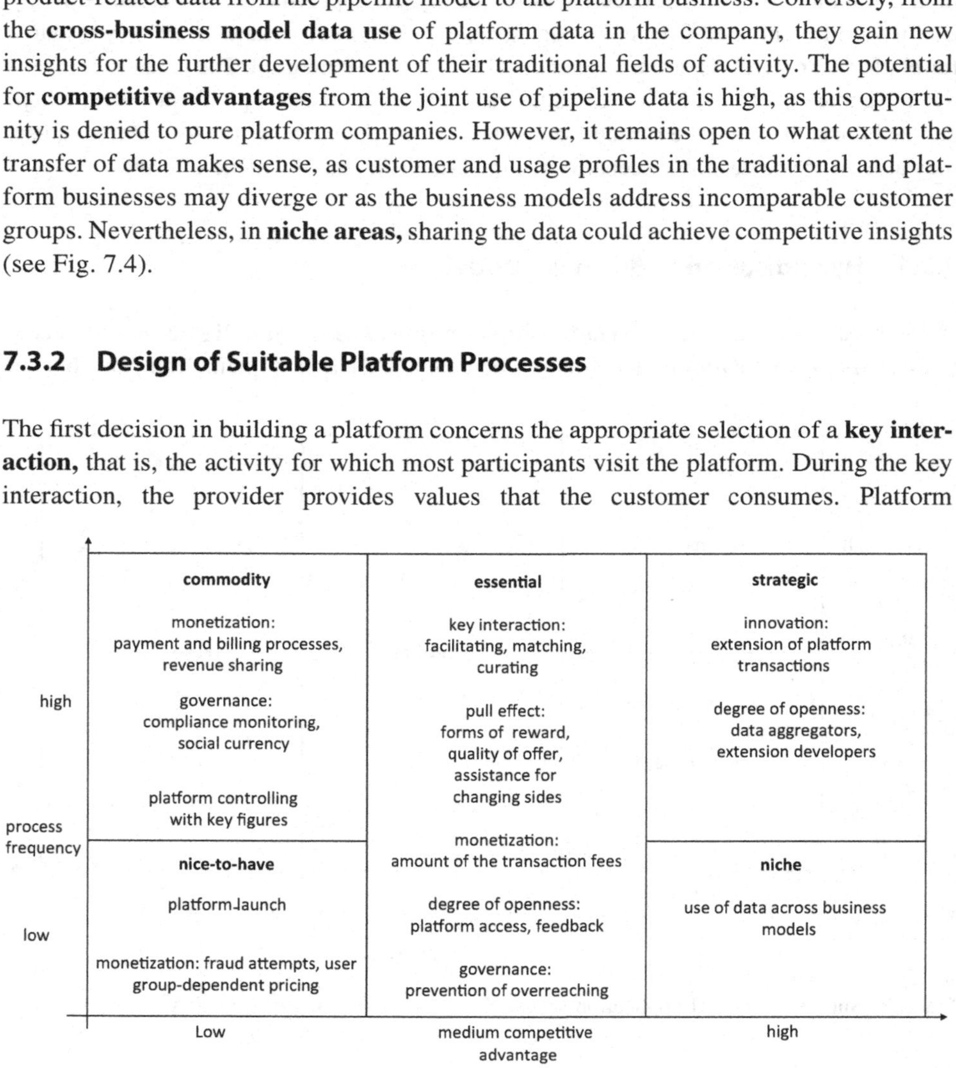

Fig. 7.4 AI application scenarios in platforms and their significance for platform success

participants exchange information, goods or services, and a currency in their interactions. It is possible that the intended value creation already occurs with the initial **information exchange.** Otherwise, the transfer of information is followed by **trade in value units** such as products or services, which in the case of digital offerings can be handled in whole or in part via the platform, while physical processes for the exchange of goods or services take place outside the platform, but may be reflected digitally on the platform in a comparable track-and-trace manner (delivery tracking). Users pay for units of value by exchanging some kind of currency, such as recognized payment instruments, cryptocurrencies, or even by transferring value in the form of ratings, attention, awareness, influence, reputation, and other intangibles. The **exchange of currency units** can be done through the platform or outside. However, controlling and verifying the payment activities of the platform helps to monetize the value creation (cf. [1, p. 45ff.]).

An important criterion for platform success is **simplicity of interactions** with a minimum of time and clicks and without the mandatory use of cumbersome tools. **Facilitating** should ensure the smooth flow of core interactions on the platform, for example, through tools for content creation, sharing and collaboration, by lowering the barriers to entry for platform use or, in cases where a lack of trust in unknown interaction partners has an inhibiting effect on users, by raising the barriers to entry (cf. [20, p. 68]). Since the network does not create the offers itself, its influence on the **quality** of the value units is low. However, on the basis of AI evaluations, the platform can provide the broadcasters with **assistance** in the creation of contributions, for example with references to successful offers or by labelling or indexing for easier retrieval by searching users (cf. [1, p. 87ff.]). Many platforms already work with ML algorithms for this purpose, such as eBay, where users receive suggestions for improving the description text, images, and appropriate pricing when they post an offer. For this purpose, the platform can use the offer category and other offer features and classify the new value unit accordingly **(supervised learning).**

Matching is another platform process critical to success and means bringing together the right users with worthwhile matches with minimal resource investment such as time or effort (cf. [20, p. 68]). A prerequisite for effective matching is the **curation** of content provided by providers in terms of its organization, selection, and presentation to other market participants. Based on the curation, a **filtering system** decides on the value units displayed to customers. The platform promotes key interactions, e.g. by supporting the provision of meaningful value units with a kind of quality control and by aligning the filter settings in such a way that customers see suitable value units while blocking uninteresting ones (cf. [1, p. 45ff.]). While simple filtering systems are sufficient for platform startups with a still small number of participants and offers, they have to evaluate and correlate a vast amount of data for large platforms. ML tools help to fine-tune the filters by deriving criteria for interesting offers from the platform's data collection, continuously monitoring the success of the proposal system, and constantly refining it. **Unsupervised learning,** for example, can use **clustering techniques to** segment users into clusters and evaluate which offers are appropriate in which group to personalize the displayed offers depending on the user. ML algorithms can also identify barriers to platform use, for example, by evaluating search patterns, typical navigation paths, and successful or aborted interactions on the platform. With the help of classification methods of **supervised learning,** it is possible to deduce which constellations lead to the termination of transactions, which tools or functionalities support the users in the best possible way, or which information about the transaction partners builds trust.

An **effective platform design** should use the **pull effect** to attract customers to the platform and keep them there. Here, the platform can build **feedback loops** with self-reinforcing activities. If a user triggers a wealth of relevant and interesting value units for himself through an interaction, he will visit the platform repeatedly to start further interactions (cf. [1, p. 54ff.]). An AI algorithm can analyze the user's activities, draw conclusions about his interests, preferences, and needs, and suggest suitable contacts and value units for further interactions. As the platform knows more about the visitor with each interaction, it can make increasingly accurate recommendations and increase the number of interactions the customer has on the platform.

Broadcasters often receive **social feedback** for their contributions on the platform, which leads to feelings of joy, awareness, recognition or wealth as rewards. As shown in Sect. 7.1.3, customer loyalty approaches are also promising for engagement with the platform as they reward activities with value units. However, financial incentives for the provision of value units should be treated with caution, are usually only effective in the short term and represent a cost burden for the platform. **Clever forms of rewards** work back on the platform itself by offering as a reward something that the platform provides internally, such as storage space, permissions, user status changes, etc. In this context, AI can help find the optimal form of reward or offer different types of rewards to different users. As a form of **unsupervised learning,** ML clustering methods can again be considered, which analyze what motivated users with similar profiles to interact more.

The key interactions with facilitating, matching and curating as well as the processes to increase the pull effect and feedback loops form the basis for platform success. Even though the process frequency may be low at the beginning, a platform can only compete if it effectively supports the platform interactions. All essential platform processes are therefore to be classified as **essential** for the competitiveness of the platform (cf. Fig. 7.4).

Innovative Platforms prove to be a very promising business model cluster in Sect. 7.1.3. Consequently, the degree of innovation of the platform is an important criterion for longer-term success. After successfully initiating the key interaction, many platforms plan to **expand the platform spectrum to include further transactions,** for example due to experience with the key interaction or the need for further growth. Opportunities for interaction expansion include changing or expanding the value units exchanged, adding a new user category (e.g., intermediaries in addition to vendors and customers), and curation to introduce new user groups, such as premium users. AI can also make an important contribution here, such as when the algorithm examines lost users and their characteristics to figure out which direction to take concerning the platform processes. By expanding their platform offerings, platform companies are trying to expand their market position and ensure their survival in the long term. Identifying suitable expansion directions is therefore a **strategic ML application field,** as shown in Fig. 7.4.

7.3.3 Platform Launch

The launch of a platform must find a solution to the **chicken-and-egg problem** that a platform is attractive to consumers if they can choose from many suitable offers, whereas the platform only attracts producers if they can hope for many demanders. Some very successful platforms, such as Amazon, have circumvented this challenge by starting as a pipeline company, in Amazon's case as a bookseller, and building a customer and supplier base in this way. If the pipeline model is successful, they redesign it into a platform, in Amazon's case this corresponds to opening Amazon Marketplace, and take users to the newly launched platform. However, this **follow-the-rabbit strategy** cannot be applied in every scenario. In addition, there are a number of other launch strategies that also do not necessarily require the use of AI (for all strategy alternatives, see [1, p. 87ff.]). Sometimes it is sufficient for a successful platform launch to **stage added value** for a user side, e.g. by paid contracting of third parties to contribute value units, or to **increase the platform attractiveness** for them, e.g. by useful tools. The **testimonial strategy assumes** that attracting a single user group is critical to platform success and specifically lures them with special benefits or rewards. The **single-side strategy** starts by building a pipeline to first acquire the participants of one side of the market as customers who will add value to the newly platform-testing users of the other side during the platform conversion. The **big bang strategy** relies on push marketing strategies to create overnight awareness of the platform and trigger a chain reaction. The **micromarket strategy** initially establishes itself in a tiny market segment, e.g. regionally limited or focused on a specific topic, and only expands once an active platform life has settled in.

Other launch strategies tend to benefit from the use of AI. The **piggyback strategy** leverages he success of an existing platform to attract its users to its own system, just as PayPal tapped into the huge user base of auction platform eBay and offered its payment services to eBay users. The piggyback strategy can use ML tools to identify suitable prospects who are presumably receptive to the platform offering by **classifying users.** In the **seeding strategy,** the platform itself participates in generating the initial units of value for one of the market sites, and thus plays a key role in determining the nature and quality of the offerings as a model for subsequent providers. The platform's push can also begin in particular through simulated, i.e. actually fake, posts. For example, in the early days PayPal used a programmed AI bot that shopped on eBay and brought sellers to PayPal. Since the bot then offered the auctioned goods directly for sale again, warehousing and shipping of goods were obsolete, while at the same time the chance increased that the new buyers also tried PayPal as a payment platform. Companies benefit here in AI use primarily from the **automation of platform processes.** Overall, AI use is not critical for platform launches. Many launch strategies do not require ML evaluations. For others, AI may be able to increase the impact of the strategy. However, the competitive advantages appear to be low overall or only effective in special constellations, so that AI can be classified as a **nice-to-have** for the platform launch (see Fig. 7.4).

7.3.4   Monetisation of Platform Services

Monetizing platform services is a major challenge, as any form of cost, such as access or usage fees, that the platform imposes on users will hinder access, dampen frequency of use, and potentially discourage platform users. Since every platform relies on revenue, it is important to provide value units for **free in some cases.** Generally, platforms charge for the exceptional value provided. The **fee type** for consumers is **access to the value generated on the platform** in the context of transactions, for providers **access to a community or market** for their value units, for both sides **access to tools and services** around the interaction as well as **curation processes** to improve the quality of the interactions (cf. [1, p. 113ff.]). Monetization via **flat-rate or revenue-based transaction fees** is suitable when a large number of similarly large transactions can be expected. They can be easily linked to payment transactions for the exchanged goods and services and have the advantage that they are only incurred when added value has been created for both sides of the market because the transaction has taken place. The level of fees perceived as appropriate must be determined by platform operators through experimentation, for example after evaluating different fee scales using an ML algorithm.

However, the platform must also ensure that the interaction of the market participants actually takes place via the platform after they have found each other. The partners might strive to make arrangements outside the platform and settle the transaction directly among themselves to save the seller the fee he shares with the buyer via a price discount. Many platforms prevent direct communication with the transaction partner. If interaction is indispensable, for example, for negotiations between the partners or for managing workflows in the course of the transaction, platforms can offer **additional added value** through **tools** and thus keep the players in the platform. Nevertheless, the danger remains that platforms for the exchange of simple services lose their users after contact has been initiated, because the service is in any case provided outside the platform, the parties involved get to know each other in person, and the customer often even supervises the provision of the service himself (cf. [1, p. 113ff.]). AI systems could infer the risk of a user not completing an interaction from scratch on the basis of past data, classify this as a suspected **fraud attempt,** and exclude the market participant from further transactions. However, this rigorous approach could also have a deterrent effect and reduce the popularity of the platform. It should therefore only be used in extreme cases and is a nice-to-have scenario.

Monetization through **fees for access to the market or community** enables providers to stand out from the crowd of rival market participants and get noticed, for example through targeted messages, better positioning as the top hit, or by promoting interactions with a particularly attractive clientele. However, demanders react angrily if they cannot distinguish between ranking positions due to paid advertising activities and high placement due to popularity that came about organically, so most platforms make both types of entries transparent and distinguish them visually. Users judge native advertising (disguised advertising), which makes paid listings look like unpaid content, as fraudulent and may avoid the platform afterwards. Other risks include a presumption of limited reach for the free content, or more lax curation of posts that have received funds for placement. Both also tend to lead to user annoyance. This is consistent with the evaluation in Sect. 7.1.3, where platforms based on the Hidden Revenue cluster tended to have poor chances of success. Overall, hardly any AI application opportunities open up for monetisation via market access.

The payment of **fees for advanced curation** only takes effect when the platform has already become so large that users find it increasingly difficult to search for high-quality content (cf. [1, p. 113ff.]). It requires AI algorithms that manage and at least partially automate curation processes, as already explained in Sect. 7.3.2.

This is followed by the question of which group of participants should pay the fees. The procedure of charging **all users** is only appropriate for platforms in exceptional cases, as it negatively affects the number of users and thus weakens network effects. Charging one side to subsidize the other is appropriate when one side of the market turns to the platform with great interest while the other remains rather indifferent. **User group-based pricing** combines subsidized free access for some users with pay models for other groups. The AI-relevant question here is which user group is allowed to operate free of charge and who is asked to pay to what extent without reducing platform attractiveness. Some platforms charge full price to most users, but subsidize stars whose presence on the platform attracts many more users. Targeting **subsidies** to price-sensitive users, who are enticed to join the platform through discounts or free offers, can redress an imbalance in market sides. However, ML-based estimation of price sensitivity can also prove difficult and may be subject to change over time. In particular, classifying a user as a star tends to be the exceptional case on the platform, so that the use of AI should be classified as **nice-to-have** (Fig. 7.4).

At launch, many platforms start free for all users to get the chain reaction of network effects going. The **freemium model** helps to attract users by offering basic versions of a product or service free of charge, while full use or extensions are only available for a fee. However, the study of business model clusters in Sect. 7.1.3 shows that freemium platforms tend to perform poorly in the comparison of success. To ensure that the **changeover** from a free platform launch to one of the monetisation variants is nevertheless successful, some restrictions must therefore be observed. For example, no added values that were previously available for free may now be subject to fees. Users also resent the sudden restriction of access to added values. On the contrary, the transition to chargeability should be accompanied by an additional new added value that justifies the price. Here, AI evaluations can provide clues as to what a meaningful added value could consist of. It is important that the design of the platform is already at the beginning in such a way that the platform has access to the values and information used for monetization. For example, capturing transaction details is essential for collecting transaction fees.

The **payment and billing processes** associated with the monetization of the platform services can be left to a rule-based billing system **(commodity)** for process optimization and automation, i.e., a rigid AI system for Automation or Assisted Intelligence (see Fig. 7.4). The billing tool autonomously implements the defined monetization strategy for a completed transaction or a fee-based additional service, determines whether fees should be charged to market participants, calculates the fees to be paid, and performs the billing. If the fee amount depends on attributes of the market participants, the transaction type or product respectively service characteristics, the algorithm can independently select and evaluate the required data. As the above-average success of the revenue sharing model shows (cf. Sect. 7.1.3), the **distribution of revenues** is also a key success criterion for the platform. In this context, the platform participants should profit according to the degree of their contributions to value creation. The platform software must therefore also organise the onward settlement to the stakeholders after the settlement has been completed.

7.3.5   Determination of a Suitable Degree of Openness

When determining the **degree of openness of a platform,** i.e. the question of which participant or partner the platform allows or prohibits which activities, platform companies walk a fine line. If they accept all content without checking it and do not control the users, they run the risk of faulty, unwanted or prohibited content. By contrast, erecting barriers to access and participation reduces the number of platform users and thus the added value of the platform. A platform is fully open if it does not restrict participants in the development, commercialisation and use of content, or if restrictions, such as the obligation to comply with technical standards or to pay royalties, are reasonable and apply without discrimination, i.e. uniformly to all users. An **open system makes** monetisation and control of contributed intellectual property more difficult, but encourages innovation. **Closed platforms** exclude or discourage individual participants through high fees or cumbersome rules. They often suffer from suboptimal processes due to the lack of a broad developer community for improvement (cf. [1, p. 136f.]). In Sect. 7.1.3, Crowdsourcing Platforms perform as a highly successful business model cluster. This type of platform benefits from opening up to third parties from their ideas and improvements, which speaks for a higher degree of openness.

One of the decisions to be made is the **involvement of external parties** who develop new value-added features for the platform over the course of the platform's lifecycle and skim off some of the added value for themselves in return. For example, **data aggregators** improve search capabilities on the platform, add more data to the platform, collect data on users and interactions, and sell it to advertisers for targeted ad placement, for example. Data aggregators and the platform itself benefit enormously from ML algorithms for their analytics, ideally creating a seamless and enjoyable user experience for platform customers, rather than being intrusive or even creepy with inappropriate suggestions. **Extension developers** integrate innovations into the platform. ML algorithms do this by capturing, for example, what add-on features developers are providing and deducing what functionality the platform is still missing. To identify particularly useful functions, the evaluation can classify, for example, which functionality several developers provide independently of each other, how often users search for an add-on, how popular the feature is among platform users, or whether the function is equally useful for different customer groups such as B2B and B2C participants or users from different industries. The platform provider can use this as an opportunity to incorporate the functionality into the standard and make it openly available. This kind of community monitoring for finding useful platform features accelerates the innovation of the platform, improves its services for all platform users and represents a strategic competitive advantage (cf. Fig. 7.4).

Furthermore, the platform must decide on the **degree of openness to users and providers** with the aim of enabling the generation of the highest quality content possible. This often slows down full openness and leads to partial openness through clever **curation,** which, for example, uses the social pressure of the community to find and remove undesirable content. On the one hand, the typical form of curation is based on **screening the access points** to the platform. However, human gatekeepers, i.e. moderators who personally screen users and content and provide feedback, are costly and time-consuming. This is remedied by AI methods for classifying interested parties on the basis of freely available or purchased data about them, which decides who should receive platform access. **Feedback mechanisms** try to positively influence the behavior of already admitted users. The Users' **reputation** is derived from their behavior on and off the platform. Special AI software can, for example, identify content providers who have already attracted attention in the past due to low-quality or inappropriate content. Tagging tools can scan posts for conspicuousness, highlighting them and alerting other users or developers to a need for revision. Similarly, AI algorithms can identify users who should be given special rights on the platform based on their positive engagement and the quality of their posts. In addition, **self-curation** or feedback from other platform participants can be incorporated by the algorithms into the classification process as a basis for decision-making. The establishment of skilful curation is **essential** essential for success of the platform (cf. Fig. 7.4).

7.3.6   Platform Governance

Governance defines the framework conditions under which the participants in an ecosystem operate. The rules for good governance include creating added value for customers through platform transactions, refraining from abusing power to take advantage or excessively retaining added value, and instead distributing it fairly (cf. [1, p. 162]). If such rules are missing or if the platform operator does not monitor their compliance, market failures occur, leading to the absence of platform transactions or to poor interactions. Rules, software architecture and markets can be identified as **governance control systems. Rules** that apply to platform users should be clearly communicated and designed to reduce or eliminate misconduct. For example, in rating systems, it must be clear what action earns a user how many points and what additional permissions or benefits the platform grants at what score. However, too much transparency is a hindrance precisely when the platform sanctions undesirable behavior, because users then deliberately circumvent the sanction mechanism by changing their behavior slightly, for example. As a tendency, the AI processes should be set up to provide quick and open feedback for desired behavior, and rather delayed and unclear feedback when punishing bad behavior (cf. [1, p. 170f.]). The automated implementation of the set of rules follows predefined conditions and can be classified as a **commodity** (see Fig. 7.4).

The **platform's program code** and the AI algorithms within it can also work towards reasonable user behaviour, for example by predicting the presumed behaviour of market participants, such as the risk of a fraud attempt (cf. [1, p. 174ff.]). Buyers who purchase incompletely or incorrectly described products far below market value, resell them far above their purchase price, and thereby achieve **arbitrage,** although exploiting market failures, nevertheless ensure the increase of the transaction volume of the platform, since without their activity for the products no transactions would have occurred. Classification procedures can provide for the punishment of arbitrageurs through deprivation of privileges by automatically detecting arbitrage attempts. Since loss of trust deters platform users, it is **essential** for the platform to prevent or minimize overreaching (cf. Fig. 7.4).

Finally, **market mechanisms** of the platform also contribute to governance by exploiting human motivation and the pursuit of pleasure, fame and success and by incorporating subjective values such as one's own preferences or personal reputation on the platform. Platforms can design a **social currency** for this purpose, such as a points system that rewards one's contributions with points that users can exchange for platform services. In addition, the platform can reward the achievement of certain point levels through charitable engagement or environmental investments, which may be in line with users' values. Professional platform participants, or those without personal contributions, purchase platform points in exchange for fee payments. Social currency also stimulates the acquisition of feedback for platform services, such as the discovery of product defects or ideas for service improvement (cf. [1, p. 176ff.]). According to Sect. 7.1.3, social currency proves to be a good basis for platform success, as can be seen from the performance of the Customer Loyalty pattern, and is a widespread mechanism for customer loyalty (**commodity**) (cf. Fig. 7.3), which AI can support with the help of rule-based processes for awarding and redeeming currency units.

7.3.7   Key Figures for Platform Controlling

While pipeline companies monitor their success with a manageable number of standard KPIs that analyze profits along revenue and cost lines, e.g., with **metrics** such as cash flow, inventory turns, or operating income, platform companies need to measure **platform success** in a different way. In the abstract, traditional KPIs show how value or resources move through the pipeline and draw attention to bottlenecks, blockages, and disruptions. For a platform, therefore, they need to map activity on the platform, that is, the rate of satisfactory interactions and the contributing factors. In this respect, the platform metrics correlate with the network effects and quantify the favoring of consistently repeated desired interactions by including the creation, sharing, and provision of added value across the entire platform (cf. [1, p. 187ff.]).

At the **beginning of the platform lifecycle,** the focus is on active providers and customers who participate intensively in interactions. The metrics must measure liquidity and matchmaking quality. **Liquidity** describes a state in which a minimal number of suppliers and demanders perform a high percentage of successful interactions. For example, it can be measured by the percentage of offers that result in interactions within a given time period. Special attention should be paid to the unsuccessful situations and the reasons why interactions do not occur, in order to avoid them in the future. The rate of active users or their growth rate also provide information about the state of the platform. **Matchmaking quality** examines the precision of the search algorithm and the intuitiveness of the navigation tools. Examples of metrics include sales completion rate as the percentage of searches with successful interaction completion. The multitude of conceivable metrics must be tailored to the respective platform (cf. [1, p. 193ff.]).

During the **growth phase of the platform,** additional key figures come into focus, such as those that show whether the two **marketplace sides balance each other out,** such as the supplier-customer ratio, the frequency of supplier participation, the number of offers created or the successful sales closures. On the customer side, important metrics include frequency of purchases and searches, and sales close rate. Approaches to calculate the **lifetime value of** platform participants from the combination of different metrics are also promising. Since suppliers can also become demanders and vice versa, the **page switching rate** finally provides indications of the balance and flexibility of the platform pages (cf. [1, p. 198ff.]).

To control the platform, it is essential to monitor the situation of the platform using key figures. However, all key figures can be calculated formula-based from the platform's dataset with rigid AI systems for Automation or Assisted Intelligence. In this respect, platform controlling neither represents a competitive advantage nor does it require complex ML algorithms (**commodity**) (see Fig. 7.4).

7.4   Recommendations for Action

7.4.1   Further Development into a Platform Company

The transformation from pipeline company to platform requires a fundamental **redesign of the business model.** Section 7.1.3 explains promising business model patterns and clusters. New **IT technologies** promote new business models and patterns (see [21, p. 100f.]). For example, the proliferation of smartphones or 3-D printing helps to recruit new groups of providers for **value delivery processes**, as they enable content creation and low-volume manufacturing even for non-professionals. Platforms also need to rethink **value consumption processes** and take into account **new forms of customer behavior** in a share economy where the use of products, services and information no longer requires ownership of them. Crucial are **networking approaches** with external partners to build a platform ecosystem that motivate platform players to deliver new forms of value to the other side. Former competitors can also be considered as platform partners. To keep as many processes as possible on the platform, consider how data can add value to the goods and services provided. In particular, **linking data to a product** not only represents a new product, but can be the starting point for a platform if the company can use either its own data or information from the community to add value to the product.

The existentially necessary growth of a platform depends crucially on the ability to **scale the business model.** To achieve this, platforms must rely on **external resources** and drive the expansion of the platform ecosystem in the sense of crowdsourcing. In addition, successful growth requires the extensive automation of processes such as participant management, matching, transaction processing, but also the identification of potential for improvement and the development of innovation ideas through analysis of the platform data. If the market sides grow at different rates, the platform operator should take countermeasures and motivate the users of one side to also become active on the other market side. Platforms must shift quality control as a remedy against negative network effects towards community-driven curation. While manual curation by human actors is sometimes still necessary at the start of a platform, mature platforms are increasingly replacing this with **intelligent automated processes.**

However, it can be difficult to build more platforms alongside competition from already established ones. The large platform companies benefit from an overflowing **cross-domain data collection.** For example, they extend their initial e-commerce data to application areas such as smart home, healthcare, or entertainment, and gain insights from semantically connecting the domains (cf. [23]). Ultimately, platform data analytics are comparable to a **barrier to entry:** other platform providers do not have access to a platform's rich data, cannot generate a similar level of value for platform participants, and will therefore be able to attract fewer interactions to their platform, which in turn leaves them without data for analytics to improve platform design.

7.4.2   Participation in the Value Network with Independent Platform

Many companies also consider appearing as a **provider on established platforms** in order to tap into a larger target audience. However, IT technologies as enablers of the digital transformation open up direct digital customer access to the platforms (cf. [22, p. 108f. and 115f.]), so that they undermine customer relationships, sales channels and consumer brand awareness. Companies thus give up valuable direct customer contacts and have to settle for data snippets and analyses provided by the platform. The platform company itself, on the other hand, has a comprehensive view of the customer and is increasingly using AI assistants to better understand the customer, create targeted brand marketing, accurately identify stages in the shopping process, optimize the shopping process, make improved product recommendations, create a tailored shopping experience, and influence

the ultimate purchase decision. The platform company is out to perfectly satisfy the customer needs to prevent users from leaving the market and selects the most suitable offers
for the platform visitors regardless of the brand. Platform participation as a provider therefore weakens the position of the company, which becomes increasingly dependent on the
platform.

However, AI can be seen as an **opportunity** for old economy companies to stand up
to platform companies by using their years of experience to run processes autonomously in a value network, retaining their knowledge and using the data profitably for
themselves. However, since a single company usually has too little data, an **independent platform approach** can help. In this approach, companies collect all the data
they generate, forward it to an independent and secure platform that operates a data
pool for old-economy participants, receives access to the shared data collection in
return, and uses it to develop its own AI for the company that helps develop new business models, offerings, services, and so on. The data and all new experiences and
transactions move back into the pool (cf. [23]). Even if the platform AI has full data
access, the intellectual property remains under the control of the data owner, which
could be ensured by an independent third party entity running the platform. By merging with other companies, all partners reach the critical size for meaningful AI evaluations without losing their independence.

References

1. Parker GG, Van Alstyne MW, Choudary SP (2017) Die Plattform-Revolution. mitp, Frechen
2. Parker GG, Van Alstyne MW, Choudary SP (2016) Pipelines, platforms, and the new rules of
   strategy. Harv Bus Rev. https://hbr.org/2016/04/pipelines-platforms-and-the-new-rules-of-
   strategy. Accessed 08 May 2024
3. Eisenmann T et al (2007) Platform networks—core concepts. MIT Cent Dig Bus 232. https://
   www.academia.edu/27832418/Platform_Networks_Core_Concepts_Executive_Summary.
   Accessed 08 May 2024
4. Grösser S (2020) Geschäftsmodell—Definition. Gabler Wirtschaftslexikon. https://wirtschafts-
   lexikon.gabler.de/definition/geschaeftsmodell-52275. Accessed 08 May 2024
5. Gassmann O, Frankenberger K, Csik M (2017) Geschäftsmodelle entwickeln: 55 innovative
   Konzepte mit dem St. Galler Business Model Navigator, 2. Aufl. Hanser, München
6. Doleski ÔD Geschäftsmodellforum—Business Model Forum, Muster. https://www.
   geschaeftsmodell-forum.com/basis/muster/. Accessed 08 May 2024
7. BMI Lab AG, 55 Pattern Cards, St. Galler Business Model Navigator, Pattern Card 52: Two-
   sided Market
8. Böhm M et al (2017) The business model DNA: towards an approach for predicting business
   model success, conference paper. In: 13th international conference on Wirtschaftsinformatik,
   St. Gallen
9. Wichert A (2020) Künstliche Intelligenz, Lexikon der Neurowissenschaft. https://www.spe-
   ktrum.de/lexikon/neurowissenschaft/kuenstliche-intelligenz/6810. Accessed 08 May 2024
10. o. A. (2020) Künstliche Intelligenz. https://www.gruenderszene.de/lexikon/begriffe/kuenstliche-
    intelligenz?interstitial. Accessed 08 May 2024

11. Frochte J (2018) Maschinelles Lernen. Hanser, München
12. Wikipedia (2020) Künstliche Intelligenz. https://de.wikipedia.org/wiki/K%C3%BCnstliche_Intelligenz. Accessed 08 May 2024
13. Futurezone (2018) Deshalb sind künstliche Intelligenz und Maschinelles Lernen nicht dasselbe. https://www.futurezone.de/science/article214871107/Deshalb-sind-kuenstliche-Intelligenz-und-maschinelles-Lernen-nicht-dasselbe.html. Accessed 08 May 2024
14. Wikipedia (2020) Überwachtes Lernen und Unüberwachtes Lernen. https://de.wikipedia.org/wiki/%C3%9Cberwachtes_Lernen#Siehe_auch und https://de.wikipedia.org/wiki/Un%C3%BCberwachtes_Lernen. Accessed 08 May 2024
15. Rao AS, Verweij G (2017) Sizing the prize: what's the real value of AI for your business and how can you capitalise? In: PwC Publication, PwC. https://www.pwc.com/gx/en/issues/analytics/assets/pwc-ai-analysis-sizing-the-prize-report.pdf. Accessed 16 März 2020
16. Weizenbaum Institut (2018) Berliner Start-ups: Aufstieg datenbasierter Geschäftsmodelle. https://www.weizenbaum-institut.de/news/detail/berliner-start-ups-aufstieg-datenbasierter-geschftsmodelle/. Accessed 08 May 2024
17. Plattform Lernende Systeme (Hrsg) (2019) Neue Geschäftsmodelle mit Künstlicher Intelligenz—Bericht der Arbeitsgruppe Geschäftsmodellinnovationen, München. https://www.plattform-lernende-systeme.de/files/Downloads/Publikationen/AG4_Bericht_231019.pdf. Accessed 08 May 2024
18. Gillham J et al (2018) The macroeconomic impact of artificial intelligence. PricewaterhouseCoopers. https://www.pwc.co.uk/economic-services/assets/macroeconomic-impact-of-ai-technical-report-feb-18.pdf. Accessed 08 May 2024
19. Michel U (2017) Controlling digitaler Geschäftsmodelle. In: Kieninger M (Hrsg) Digitalisierung der Unternehmenssteuerung. Schäffer-Pöschel, Stuttgart, p. 33–50
20. Jaekel M (2017) Die Macht der digitalen Plattformen. Springer Vieweg, Wiesbaden
21. Roth-Dietrich G, Gröschel M (2018a) Matching zwischen innovativen Geschäftsmodellmustern und IT-Wirkungsbereichen. In: Barton M, Müller Ch, Seel Ch (Hrsg) Digitalisierung in Unternehmen. Springer Vieweg, Wiesbaden, p. 87–109
22. Roth-Dietrich G, Gröschel M (2018b) Digitale Transformation: Herausforderung für das Geschäftsmodell und Rolle der IT. In: Lang M (Hrsg) IT-Management. DeGruyter, Oldenbourg, p. 97–122
23. Büst R (2018) Künstliche Intelligenz: In 7 Schritten zum AI-enabled Enterprise, Computerwoche Online. https://www.computerwoche.de/a/in-sieben-schritten-zum-ai-enabled-enterprise,3544401. Accessed 08 May 2024

Prof. Dr. Gabriele Roth-Dietrich holds a degree in physics and a doctorate in business administration from the University of Mannheim on process optimization and automation in healthcare. She worked for almost 10 years as a project manager and system analyst in development and product management at SAP SE. After a professorship at Heilbronn University, she has been teaching business informatics at Mannheim University of Applied Sciences since 2011, focusing on enterprise software, workflow management, business intelligence, project management and digital transformation. E-mail: g.roth-dietrich@hs-mannheim.de. Web: https://www.informatik.hs-mannheim.de/wir/menschen/professoren/prof-dr-gabriele-roth-dietrich.html.

Part IV

Operational Scenarios

Artificial Intelligence in Automated Document Processing Using the Example of Health Insurance Companies

8

Gerhard Hausmann and Uwe Lämmel

Abstract

The fully automated processing of documents in insurance claims is referred as dark processing by German insurers. Dark processing is a form of digitalization in document processing. For cost reasons, insurance companies have been working for years to increase the share of dark processing. Invoices in private health insurance in Germany offer good conditions for automated processing due to the fee code system based on the German fee schedule. Both classic artificial intelligence (AI) techniques such as rule systems and newer connectionist approaches such as convolutional neural networks (CNN) with deep learning are used here.

8.1 Digitisation of Document Processing

The manual processing of documents is time-consuming and cost-intensive. As early as the 1980s, the idea of a paperless office was born, but the development of copying and printing technology turned it into its opposite. Paper documents are still received and must not only be scanned but also have their content recorded so that the data is available in the company's IT systems. This is just as true for documents sent via an invoicing app, as this only eliminates the need for scanning. Artificial intelligence (AI) techniques help to

G. Hausmann
Abt. Infrastruktur- und Architekturmanagement, Barmenia, Wuppertal, Germany
e-mail: gerhard.hausmann@barmenia.de

U. Lämmel (✉)
Hochschule Wismar, Wismar, Germany
e-mail: uwe.laemmel@hs-wismar.de

T. Barton, C. Müller (eds.), *Artificial intelligence in application*,
https://doi.org/10.1007/978-3-658-43843-2_8

"understand" the texts recognized by OCR. The aim is to capture data and recognize the main content of a text, thus enabling further processing.

For many routine processes, fully automatic processing of the process initiated by a document is desirable. This is referred to as dark processing in Germany. The Gabler Versicherungslexikon defines dark processing as: "Part of a business process within electronic document management following the scanning and classification (indexing) of documents. After the indexing by means of automated check mechanisms and process sequences, a case-final processing of a business case follows by the IT system. Due to the lack of intervention by an employee and the fact that it is not visible to them (in the 'dark'), this form of business transaction processing gets its name." [1].

Decisive for dark processing is not only the recognition of the characters of the document, but the "understanding" of the content. Structured documents are easier to analyze than pure continuous texts. The service invoices of the insured persons in the private health insurances are structured and semi-formalized documents and are thus particularly suitable for automatic processing. In the German Fee Schedule for Physicians (Gebührenordnung für Ärzte GOÄ), each medical service is assigned a key, a short description, and a monetary amount. These individual details can be compared with each other and thus enable reliable recognition and "understanding" of the individual items of a service invoice.

However, individual items cannot be billed at will. There are dependencies between the individual services as well as dependencies on the insured person's tariff. A fully automatic processing of a service accounting (a dark processing) from the incoming service accounting to the payment order must recognize both the line items and the dependencies and take them into account accordingly. Such dependencies can be described well using rules based on logic. The prerequisite for this is that the details of an invoice are recognized as unambiguously as possible. For this recognition, which should be as free of doubt as possible, connectionist approaches such as Convolutional Neural Networks (CNN) can be used, which are trained using Deep Learning.

The use of AI techniques is not limited to such service billing. The "understanding" of texts can be used in general to automatically include data in IT systems and thus trigger processes in a targeted manner. The areas of application are diverse, from the recording of invoices in general, the processing of certificates of incapacity to work or letters of termination, claims notifications in the area of motor vehicle insurance to the analysis of customer opinions.

8.2   Business Rules

Knowledge representation and processing using rules is one of the classical methods of Artificial Intelligence. Based on the logical connection of implication in predicate logic, relationships are formulated in IF-THEN statements: If an employee is responsible, but this employee is sick or on vacation, then his substitute is responsible. Somewhat more formally noted as a rule:

IF responsible(X) AND (sick(X) OR vacation(X)) AND substitute(X,Y) THEN responsible(Y).

By means of so-called predicates, properties (in the example responsible as well as sick) and relationships (representation) are described, which are practically stored in databases. Business rules go beyond purely logical rules like the example mentioned: One speaks of production rules, since actions are also allowed in the then-part. One also speaks of the action part of a rule:

IF processing_completed THEN send_message.

Rules are processed in such a way that the existing data (facts) are used to check which rules are applicable. A rule is applicable if the condition part (the if part) of the rule is fulfilled, i.e. the check of the logical expression results in the value "true". If several rules are applicable at one time, this is referred to as a conflict set, from which a rule is then selected and executed.

The practical use of business rules is such that knowledge about processes or decisions is represented by such rules. This typically results in a very large number of rules, several thousand, ten thousand or one hundred thousand. These are managed and processed by Business Rules Management Systems (BRMS). The advantage: Requests for changes can be implemented more easily and quickly by changing the rules than by reprogramming software parts, see Fig. 8.1.

BRMSs from various vendors are available for the development of rule-based solutions, such as Actico Rules (formerly Visual Rules for Finance), FICO® Blaze Advisor® Decision Rules Management System, IBM Operational Decision Manager, Oracle Business Rules, Red Hat JBoss BRMS, SAP Cloud Platform Business Rules. The open source solution Drools (Red Hat) extends Java with a rules language and uses the Eclipse IDE.

Fig. 8.1 Use of a rule system—Business Rules Management System (BRMS)

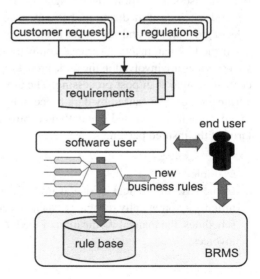

8.3   Rule Processing for Medical Invoices

Rule processing is an approach from symbol-processing AI and requires that corresponding data are available. In private health insurance, these are nominal values such as tariff keys, fee codes, diagnosis keys, keys for hospital flat rates or drug identification numbers. In addition, numeric values such as amounts, factors, age or time specifications are also taken into account.

Health insurance companies operate semi-automated data capture systems for classifying documents and extracting structured data from document images. The data initially recognized automatically by these systems is subject to errors, so that manual verification of the data is necessary, thus incurring considerable costs. If the data is then available in a structured form, rule systems for the dark processing of benefit applications can automate a wide range of checks and tasks. Two examples are considered in the following:

1. Identify the tariffs or contract statuses of the insured person that are decisive for the reimbursement of the respective invoice amount: The document type, date of treatment and tariff must be taken into account. For the reimbursement of benefits for homeopathic medical treatment, for example, a basic tariff for outpatient and inpatient medical treatment and perhaps a supplementary tariff for naturopathy must be taken into account.
2. Identify invoices or even combinations of documents in benefit claims that are to be excluded from dark processing. These include complicated claims that require the consideration of medical reports, or medical or dental invoices that violate fee schedule regulations. Rule-based decisions are also used to split benefit claims into subprocesses, some of which are processed in the background, while others are routed to benefits processing or the expert check for manual processing in the foreground.

The expertise to be modeled for dark processing is extensive: it includes provisions of tariffs, risk assessments for medical diagnoses, specifics of medical or dental services, properties and specifics of medications, basic medical knowledge, departmental requirements for selecting documents and claims for manual processing by case handlers, and much more.

At the German health insurance company Barmenia Krankenversicherung, four decision services are involved in the dark-processing process: The task of the first service is to control manual scan post-processing. The aim is to keep additional costs for the manual verification of data caused by dark processing as low as possible. Which parts of the data of a document are verified manually is controlled by rules, for example depending on the tariff of the insured person.

Example
In the case of medical or dental invoices where the rule-based verification of the initially automatically recognized data detects violations of the regulations of the fee schedules, the manual verification of the fee codes can be limited to the conspicuous invoices.

For the verification of medical and dental invoices, two independent expert systems are used, which were initially developed for manual invoice verification. For the dark processing they are extended by technical interfaces. The goal is to recognize medical and dental invoices of low to medium complexity that also do not show any abnormalities and are suitable for dark processing if all other prerequisites are met.

Finally, the task of the fourth rule system is to select the invoices and other documents that meet all the requirements for dark processing. In addition, this service determines transactions that can be fully or partially processed in the dark. Another task in the dark processing of transactions with medical or dental invoices is, for example, the recognition of invoices with exclusively prophylactic medical services, which are to be given special consideration in some tariffs.

Before rule-based decisions are made, service descriptions from medical and dental invoices are first sent to neural classifiers for texts in order to improve the data quality for the charge codes that have not been verified manually (see Sect. 8.5). The improved data is then enriched with supplementary information. This includes the contract data of the insured person, information on the tariffs, in the case of fee codes information on the associated medical services or in the case of medical diagnoses supplementary risk assessments. This knowledge is predominantly mapped in tables and managed in databases.

The technical basis of the rule systems of Barmenia Krankenversicherung is the IBM Operational Decision Manager (ODM). The architecture of ODM is a distributed architecture based on Java Enterprise Edition. Data to be considered in the decisions is modeled in an object-oriented way. The rule engine uses the get methods of the objects in the if part of the rules to access data. The properties of the objects correspond to the predicates of the predicate logic:

X.isTariffForNaturopathy() is an example of a one-character predicate,

X.isCombinableWithTariff(Y) is an example of a two-character predicate.

In the then or action part of the rule, the rule engine accesses the set methods of the objects to record decisions by changing properties of the objects. Business rules management systems (BRMS) such as ODM also allow more complex processing to be triggered during the execution of the rules via methods of the objects, for example, logic can be combined with other techniques for making decisions.

Example
Combining different methods for making decisions makes sense if the logic available for this allows several solutions. An example from invoice verification: Item A cannot be calculated next to item B, and B cannot be calculated next to C. How do you decide when an invoice lists A, B and C? If there are no other rules, there are two alternatives to choose from: Either pay A and C, or pay only item B. One strategy for automating such decisions is to evaluate the possible solutions, for example, based on the amount to be paid.

For the development of rules, ODM has an IDE that provides both an intuitive, easy-to-use rule editor and tools and utilities for working with rule projects. The executable classes of the data model, their properties and methods are first verbalized before the rules are developed, and names from the user's technical language are assigned to them. This creates a vocabulary with terms such as *"the tariff"*, *"the dentist's invoice"*, *"the amount of the invoice"* or *"the fee code of the invoice item"*.

In the rule editor, the IF-THEN rules, the business rules, are developed from these vocabulary using logical operators such as NOT, AND and OR. It is possible to define variables for objects of classes of the data model: *"P denote an invoice item"*. Quantors typical of predicate logic such as *"THERE IS NO invoice item with property ..."* or *"THERE IS AT LEAST ONE invoice item with property ..."* are also part of the rule language of ODM. Variables and quantifiers completely eliminate the need to write source code for iterating over sets or lists of objects. The rule engine ensures that all rules check all matching objects in the set of case data, see Fig. 8.2.

Extensive rule sets or decision tables can be structured and managed in packages. The order in which the rule packages are processed by the rule engine is specified by a rule flow. Like a program flow chart, the rule flow can have branches that are linked with conditions. In this way, objects for medical invoices can run through different parts of a set of rules than objects for dental invoices.

The rule-based decisions are executed on the ODM Rule Execution Server, which manages versioned rule sets and rule engines for the execution of the rules. Data models and rule sets are exported from the IDE and made available on the Rule Execution Server. Objects are transferred to the Rule Execution Server for execution, which map the case-specific information, supplemented by the name and version number of the set of rules.

To execute the rules, ODM has three types of algorithms: One of the algorithms is a forward-chaining algorithm [2], which applies the rules to the case-specific data until no further rule is applicable, that is, until no new knowledge derivable from data and rules

```
definitions:
        set 'line_item' to a line item;
if      invoice - is invoice for outpatient medical treatment
        and line item - has a fee code of the fee schedule for physicians
        and     (       line item - the fee code is one of {'30', '31', '269a', '558' }
                        or line item - the fee code (integer value) is between 284 and 286
                )
then
        invoice - log 'fee code for naturopathy'
        invoice - add attribut 'dark processing' with value 14;
```

Fig. 8.2 Business rule for the dark processing of an outpatient physician invoice. The rule identifies invoices in which naturopathy is billed on the basis of fee codes of the German physician fee schedule

emerges. This may be necessary when chains of logical inferences from multiple rules must be executed to make decisions. Often, however, the business logic is simpler and only requires the execution of single rules. Two other algorithms therefore do not use forward chaining, but are optimized for efficient use of CPU or memory. The most suitable algorithm can be individually assigned to each rule package.

The weakness of such rule systems compared to machine learning systems is that data models and business logic for dark processing have to be implemented by software and rule developers. However, the declarative approach of mapping business logic into rules and using a rule engine to process the rules greatly simplifies the implementation and maintenance of extensive or complex business logic compared to imperative programming languages. The expressions in the rule language are thus free of nested loops, which significantly increases their readability and hence their maintainability. The simple, descriptive approach combined with easy-to-use rule editors also requires little training.

One of the strengths of rule systems compared to machine learning systems is the good traceability of decisions. One of the best practices in projects with extensive business logic is to let the rule "firing" in each case generate hints, which are collected per case. To do this, supplementary actions are included in the action part of the rules. When benefit claims are processed in the dark, such notices can be included in the benefit notices that are sent to customers. This documentation of decisions supports both quality assurance in the test phase and the transparency of decisions in production operation. It thus ensures acceptance of the automated decisions by customers, by users in customer support, and by the clients in the specialist department.

8.4 Artificial Neural Networks and Deep Learning

Although the first ideas for artificial neural networks (ANNs) date back to the early days of AI in the 1950s, a breakthrough did not begin until the 1990s. Many small units are interconnected according to the model of a natural neural network and trained with mathematically based learning methods.

Figure 8.3 shows a feed-forward network as it is used in data mining for the classification of data or also for a forecast. For the application of neural networks, data sets must be available: Data from the past for which the classification or prediction result is known. The network is trained with this data and can then be applied to new data. The basic prerequisite for this is the assumption that the future will behave similarly to the past, which has been learned from.

The term Deep Learning is associated with so-called Convolutional Neural Networks. These are also feed-forward networks that contain several or even many intermediate layers, each of which is usually composed of a convolutional layer and a pooling layer. At the end, a normal, fully connected neural network is attached, see Fig. 8.4.

At the beginning of the 2010s, the success story of Convolutional Neural Networks (CNN), which are trained using Deep Learning, began. In addition to successes in

Fig. 8.3 Feed-forward
neural network

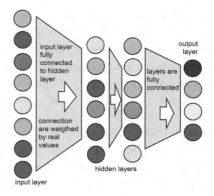

character recognition, especially handwriting recognition, CNN are successfully used in object recognition. Unlike simple neural networks, CNN can recognize objects regardless of their location in the input image. More details on how a CNN works can be found e.g. in Schwaiger and Steinwender [3].

8.5 Deep Learning in Text Processing

The machine processing of texts is a subfield of computational linguistics (Natural Language Processing, NLP). NLP encompasses a whole range of methods. These include the classification of texts, the recognition of moods expressed in texts (sentiment analysis), the recognition of names of entities such as persons, organizations or places, or machine translation. In the field of NLP, methods based on Deep Learning have increasingly been replacing older techniques for several years. Google, for example, has been using artificial neural networks for machine translation since the end of 2016 [4].

Machine classification of texts can effectively support dark processing of benefit claims in private health insurance. Use cases are:

- Classification of written notices submitted with claims, for example, when claims are submitted by photographing and sending invoices using a billing app. Does the free-form note that the customer has entered into the app interface require the control to be switched to manual processing or can it still be processed in the dark (automatically)?
- Classification of short texts from invoices or cost estimates that describe services provided by physicians or dentists. In the case of private medical treatment in Germany, fee schedules are binding for the billing of medical or dental services. The task of the classification is to assign appropriate numbers of the respective fee schedule to the short texts from the invoices. The fee codes are symbols for classes to which the services are to be assigned. Currently available systems for semi-automatic data capture in input management have only limited success in this task and require costly manual post-processing of the fee codes recognized with OCR.

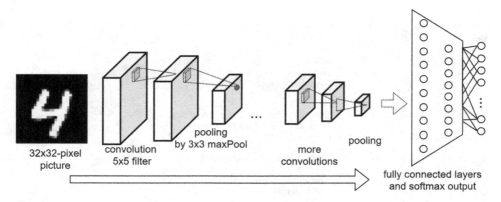

Fig. 8.4 Structure of a Convolutional Neural Network (CNN)

Once the texts from the invoice tables have been assigned suitable fee codes of sufficient quality, important decisions can be made automatically with the help of rule systems for the dark processing of documents. In this way, invoices for medical or dental services can be identified that conform to the various rules of the fee schedules and are thus suitable for dark processing.

Invoices that only contain prophylactic services such as examinations or vaccinations can also be identified on the basis of the fee codes. Private health insurance tariffs in the German insurance market often provide for premium refunds for insured persons without benefits. Costs for preventive health care are an exception and do not prevent premium refunds. For this, they must be recognized and booked in the dark processing other than treatment costs caused by illness.

A technical feature of these classifications is that many classes have to be distinguished in some of the described tasks. The fee schedule for dentists in Germany comprises 215 different fee codes, the fee schedule for physicians 2842. The results that can be achieved by applying the Deep Learning methods for the classification of short texts from tables in invoices prove to be so good, despite the many classes to be distinguished, that for the dark processing of medical or dental invoices the time-consuming, manual verification of the data of the table items recognized in the incoming mail can be dispensed with. The system for semi-automatic data entry only has to recognize and separate rows and columns of the tables with invoice items semi-reliably (Fig. 8.5).

Convolutional Neural Networks (CNN) have become the dominant technique in the field of digital image processing, but they are equally suitable for the classification of texts. For this purpose, texts are converted into two-dimensional matrices of numerical values, which are interpreted as graphics, text images or patterns in general and used as input for a CNN.

In 2015, Zhang et al. developed a "character level" method [5]. LeCun is one of the three scientists who received the 2018 Turing Award for their work in Deep Learning. In the method, a character set or "alphabet" is formed from characters commonly used in

a

Beh. dat.	Birth no.	Description	Factor	EUR
07.08.21	1	Advice, also by telephone at separate times, telephone advice to clarify the Symptoms before home visit	2,30	10,72
07.08.21	D	Surcharge for services rendered on Saturdays, Sundays or public holidays.	1,00	12,82
07.08.21	50	Visit, including consultation and symptom-related examination above-average time expenditure Child-friendly approach	3,50	65,28
07.08.21	H	Surcharge for services provided on Saturdays, Sundays or public holidays	1,00	19,82
07.08.21	E	Surcharge for urgently requested and immediate execution	1,00	9,33
07.08.21	7	Examination of organ system Difficult circumstances at home	3,50	32,64

b

Fee code (OCR)	Score	Fee code (NLP)	Score	Text
1	100	1	100	Consultation, also by telephone at separate times, telephone consultation to clarify the symptoms before the home visit
D	100	D	88	Surcharge for services provided on Saturdays, Sundays or public holidays
50	100	50	100	Visit, including counseling and symptom-related examinations ü Above-average amount of time spent Child-friendly approach
H	100	D	69	Surcharge for services rendered on Saturdays, Sundays or public holidays
	0	E	93	Surcharge for urgently requested and immediate execution
7	100	7	100	Examination of organ system Difficult circumstances in the home

Fig. 8.5 Section of a doctor's invoice in German language and table of fee codes with confidence scores recognized in input management on the basis of OCR. The data were not verified manually (grey background). The fee code "E1" in the fifth position was not recognized. The classifier for short texts (NLP), however, can correctly assign the fee code with a score of 93%. The fee codes "H" and "D" for the fourth position are synonym

texts. Each character in the text to be classified is then replaced by a vector whose length is equal to the length of the alphabet and which is predominantly filled with zeros. Only at the position corresponding to the character's position in the alphabet is a one entered: one-hot encoding. Rare characters that are not included in the alphabet can, for example, be represented by a zero vector.

Arranged side by side, these vectors result in a matrix that resembles Braille, and is suitable as input for image-processing methods. Since CNN processes input with fixed dimensions, a maximum text length is specified. Shorter texts are padded to this length with spaces, while longer texts are reduced to the maximum length. For good results, the text length must be adapted to the task so that, on the one hand, as little information as possible is lost by trimming the texts and, on the other hand, no unnecessarily long sequences of blanks are generated that are free of information and would thus adversely affect the machine learning process (Fig. 8.6).

On the one hand, character-level methods have the advantage that the preparation of the texts as input for the CNN requires very little effort. On the other hand, good results can also be achieved with texts containing errors, for example if individual characters are not correctly recognized by the OCR.

CNNs used for text classification employ one-dimensional convolutions and one-dimensional poolings, which, unlike digital image processing, move in only one direction over the encoded text. They are also referred to as temporal convolutions.

They are matrices of numerical values that are placed over the encoded text and moved character by character. At each step, the products of the superimposed numbers are added

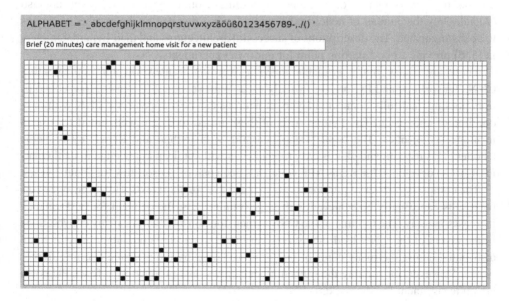

Fig. 8.6 One-hot encoding for preparing a text as input for a Convolutional Neural Network

together to form a single number. The values of these sequences of numbers thus depend on the convolution and on the combination of characters. Convolutions in the upper layer of the network learn during training to produce high activations for certain combinations of characters, but low activations for others, thus acting like filters. Combinations of combinations of characters are then processed in the layers below.

Besides the architecture with six layers of convolutions studied in 2015 at New York University, even better results can be obtained in some use cases with very deep architectures with, for example, 17 or 29 layers of convolutions [6]. The different architectures of CNN differ not only in the number of consecutive layers, but also in the width of the convolutions as well as in the factor of poolings. For very deep architectures, "narrow" convolutions with a width of only three characters and minimal pool sizes are used.

Which CNN architecture leads to good results for a specific task must be tested on a case-by-case basis. Experience shows: For very short texts with a length of often only a few words or a maximum of about 128 characters, CNNs with only exactly one layer with convolution followed by pooling and three fully networked layers achieve good results. If large training sets are available, e.g., 100,000 records or more, deeper architectures with small filter sizes and multiple minimal poolings can lead to better results. Also for the deeper architectures, if the text is rather short, the number of layers should be reduced compared to the CNNs from [6], which were constructed for a text length of 1024 characters. Non-sequential architectures with two or more paths with layers of convolution could further improve the results for the described tasks.

In addition to the number and arrangement of layers, so-called hyperparameters such as the number of convolutions per layer or the number of neurons in the two hidden layers of the classifier following the convolutions and poolings from fully connected layers are also crucial. Only the number of nodes in the output layer is dictated by the number of classes to be distinguished. The descriptions of successful architectures and their hyperparameters in [5, 6] provide a good starting point for our own experiments.

There are a few freely available machine learning frameworks for implementing CNN. TensorFlow and PyTorch are widely used. With version 2 of TensorFlow, convolution, pooling, and fully connected layers of neurons can be mapped for a CNN with only one line of code each. The Python APIs of these frameworks are well suited for use in insurance companies. The book "Deep Learning with Python and Keras" by Francois Chollet [7] is recommended, it is easy to read even without deep knowledge in mathematics. Chollet developed the Keras API, which has been part of TensorFlow for about 2 years.

The computing power requirements are manageable for short texts. However, since better results can usually be achieved with large training sets, the demand for performance increases quickly. The training of the CNN is then done on a GPU. The many convolutions in the layers of the CNN lend themselves very well to parallel computations, which are distributed across the many cores of one or more GPUs by the machine learning frameworks. However, for running the trained models in test and production, a few CPU cores are quite sufficient for short texts.

Table 8.1 Classifiers for short texts for use in dark processing and invoice verification

Classifier	Text length	Classes	Convolutions	Training and test data sets	Accuracy (%)
Classification of short user comments	96	2	3	15,000	93.00
Legends from dental laboratory bills	96	261	3	180,000	94.85
Legends from dental bills	154	195	3	1,100,000	98.14
Legends from medical bills	128	1647	11	12,500,000	98.30

Accuracy of different classifiers for short texts, which were developed at Barmenia Krankenversicherung for use in dark processing and invoice verification

Private health insurance companies in the German market such as Barmenia have been operating rule-based expert systems for checking invoices for years. Texts with assigned charge numbers are taken from their databases for training the neural networks. Machine learning based on data in conjunction with expected results is called supervised learning. The available data is often divided into training and testing data in a 4:1 ratio.

It is challenging that certain medical services occur very frequently, while others occur very rarely. The classification of free-form texts does not work satisfactorily if only a few examples are available for training. For this reason, texts for infrequently listed fee codes are combined into a single class. The CNN then learns to distinguish only those fee codes that are present with a certain minimum frequency in the training set. This is more than sufficient for practical dark processing applications that target simple, typical bills.

The measure of the quality of the classifiers is the accuracy, which is the proportion of the test data that is correctly classified by the classifier following training. To get a more accurate picture, if many classes are distinguished, the accuracy for the texts of each class can also be very informative. In case of unbalanced training sets, the accuracy of the frequently represented classes will usually show the best values, cf. Table 8.1.

For the integration of the trained classifiers into the technically heterogeneous infrastructure of a company, the classifiers can be operated as a microservice in a Python web server and encapsulated in a virtual machine or a container. The exchange of data with the service consumers then takes place via a REST API.

8.6   Apply AI Methods in an Integrated Way

AI methods have been successfully used for many years to automate processes in document processing. The examples given show that both symbol-processing AI in the form of business rule systems and the connectionist AI of artificial neural networks contribute to success and thus steadily increase the proportion of fully automatically processed documents, in this case benefit applications.

This dark processing relieves the employees of the benefits and claims departments in insurance companies, leads to cost reductions and to an increase in productivity. The processes of this document processing are slimmed down and standardized, while also being accelerated. Fast customer communication combined with high processing quality generates satisfied customers and also increases the satisfaction of the employees involved in the process.

As of December 2019, approximately 70% of prescriptions, 73% of medical invoices and 45% of dental invoices were processed in the dark at Barmenia Krankenversicherung. Completely dark processed are 49% of the benefit applications, if the share of dark processed partial processes is added, it is about 54%. An enormous saving in manual activities.

In addition to the automated processing of benefit claims, applications have been reported in the processing, or at least the automatic recognition and transfer of data into IT systems in the area of notices, claims or medical reports [8]. AI helps to understand documents and transfer the relevant data they contain into the systems. Structured documents, such as business letters for frequently recurring tasks, are a good starting point here. Complete dark processing for largely free-form documents is still difficult to imagine. A well-functioning OCR at the start of the process can make a lot of things easier, but it should also be noted that not all documents are perfectly scanned. Downstream processing using AI techniques can compensate for OCR errors and thus increase the rate of automatic processing.

References

1. Oletzky T (2017) Dunkelverarbeitung. Gabler Versicherungslexikon. Gabler. https://www.versicherungsmagazin.de/lexikon/dunkelverarbeitung-1945013.html. Accessed 10 May 2024
2. Lämmel U, Cleve J (2023) Künstliche Intelligenz, 6. Aufl. Hanser, München
3. Aggarwal C (2018) Neural networks and deep learning. Springer, Cham
4. Le QV, Schuster M (2016) A neural network for machine translation, at production scale. https://ai.googleblog.com/2016/09/a-neural-network-for-machine.html. Accessed 10 May 2024
5. Zhang X, Zhao J, LeCun Y (2015) Character-level convolutional networks for text classification. https://arxiv.org/abs/1509.01626. Accessed 10 May 2024
6. Conneau A et al. (2016) Very deep convolutional networks for text classification. https://arxiv.org/abs/1606.01781. Accessed 10 May 2024
7. Chollet F (2018) Deep learning mit Python und Keras: Das Praxis-Handbuch vom Entwickler der Keras-Bibliothek. Mitp Verlag, Frechen
8. ExB (2020) KI für Versicherungen. https://www.exb.de/versicherung/. Accessed 10 May 2024

Gerhard Hausmann After studying mathematics at the University of Wuppertal, Gerhard Hausmann first worked as a teacher in the field of vocational education. From 2000, he developed software for Barmenia, where he now works as an architect for artificial intelligence systems. His work focuses on the development of expert systems for the verification of invoices and of automata for the dark processing of benefit claims in health insurance.

Prof. Dr. Uwe Lämmel studied mathematics at the University of Rostock and received his doctorate in 1985 for his work on translator-generating systems. In 1996, he was appointed to the Wismar University of Applied Sciences for Fundamentals of Computer Science and Artificial Intelligence (AI). His research focuses on the use of AI methods in business applications, not only rule-based systems, but also in particular the use of artificial neural networks.

AI in Recruiting: Potentials, Status Quo, and Pilot Projects in Germany

Stephan Böhm, Olena Linnyk, Wolfgang Jäger, and Ingolf Teetz

Abstract

In recruiting, large amounts of data must be processed and analyzed. There is a long tradition of implementing information technology to support the recruiting process before, during, and after a job application. IT systems improve decision-making through efficient and fast processing of the data. In addition, modern communication technology can foster an efficient exchange of information between recruiters, departments, and applicants. Artificial intelligence (AI) has the potential to increase further this data processing efficiency in many fields along the "candidate journey." This chapter discusses AI's application areas and potential in recruiting. Moreover, experiences with two AI-based pilot recruiting projects in Germany are described. The first pilot is an AI-based augmented writing tool to improve job advertisements. The second pilot is on a FAQ chatbot as an extension of an applicant tracking system.

S. Böhm (✉)
RheinMain University of Applied Sciences, Wiesbaden, Germany
e-mail: stephan.boehm@hs-rm.de

O. Linnyk
milch & zucker AG, Gießen, Giessen, Germany

Frankfurt Institute for Advanced Studies, Frankfurt am Main, Germany

W. Jäger
Dr. Jäger Management-Beratung, Königstein, Germany

I. Teetz
milch & zucker AG, Gießen, Giessen, Germany

T. Barton, C. Müller (eds.), *Artificial intelligence in application*,
https://doi.org/10.1007/978-3-658-43843-2_9

9.1   Recruiting as a Field of Application for AI

Recruiting tasks can be mapped into two areas of activity: (1) "In front of the scenes," i.e., all activities that are geared towards candidates or applicants and run along the "candidate journey," and (2) "Behind the scenes," i.e., all internal processes within the company that are to be managed and performed by the recruiting organization—from the provision or approval of positions to job postings, selection processes, contract negotiation, and onboarding.

During the last 25 years, both sides have received growing digital support, not least due to the triumphant advance of the Internet, both in external personnel marketing and in internal applicant management. The internal support requirements were initially characterized by rationalization effects for large numbers of applicants. These requirements were fulfilled through applicant tracking systems and applications' associated digital, online processing. In external personnel marketing, on the other hand, the company's career websites and cost-effective new media such as job advertisements on online job boards filled the top of the agenda.

The demand for recruiting professionals has increased significantly in recent years. The need for new employees has risen strongly in almost all areas of business and administration. As a result, talent competition has intensified dramatically. In 2018, 79 percent of all vacancies that companies reported to the German Federal Employment Agency were for jobs in shortage occupations (see [1]). For example, there is a shortage of much-needed workers in STEM-related professions and the healthcare sector. At the end of April 2019, just under half a million jobs with STEM backgrounds were unfilled in Germany. By 2021, Germany will need an average of 258,600 new STEM professionals annually to maintain the current workforce [2].

Last but not least, many potential applicants are aware of their strengthened position in the recruiting market, thus increasing their expectations of the potential employers in terms of approach and interaction across the steps of the candidate journey. But pressures on recruiting departments are also increasing within their organizations. Recruitment processes are taking too long, and the number of vacancies that cannot be filled is growing. In addition, expenditure on personnel recruitment continues to rise. To cope with these challenges, improving the speed and accuracy of each process step is necessary.

The demanded and required more comprehensive technical support can only be achieved through a more consistent digitalization. AI-based solutions have particular potential in this regard. Applicants and recruiters alike can benefit from the use of AI. Objectives are, for example, more outreach with less overhead spending when addressing applicants. AI can also achieve a better candidate experience in the recruiting process and accelerate successful job placements. In addition, the efficiency of human resources should be improved through AI use. At the same time, AI deployment must comply with all relevant legal standards and laws. However, adjustments to work organization are also necessary to exploit the potential of digitization. In particular, new, more agile forms of collaboration and leadership are essential for successful digitization. Using AI in recruiting thus places diverse demands on all parties involved. However, if implementation remains stuck at the level of mere marketing promises, digitization measures will ultimately come to nothing. A high risk of failure of an AI deployment also exists if the concerns or fears of applicants and recruiters are not taken seriously. Unlawful or unethical decisions through the incorrect or ill-considered use of AI not only carry the risk of prosecution and penalties but also a loss of trust and reputation concerning the employer brand.

9.2 Special Requirements for the Use of AI in Recruiting

The term "artificial intelligence" is currently being used in an inflationary manner. AI generally refers to a scientific discipline that uses computer-based methods that enable automated problem solving modeled on human intelligence (see also [3]). This includes machine learning methods that emulate human learning processes through adaptive algorithms to identify data patterns or classify them. In this area of AI, significant progress has been made in recent years with so-called deep learning algorithms [4]. Deep learning uses complex and multilayer neural networks for machine learning. However, corresponding terms are often used only from a marketing perspective. AI terms are then introduced to promote data-based process improvements or automation, often based on improved traditional data analysis methods. Special requirements and challenges arise from applying "real" advanced AI methods in recruiting, which are described in more detail below.

9.2.1 Solution Orientation

AI-based self-learning systems do not make decisions based on examining specific predefined rules but "learn" a particular decision-making behavior based on examples in training data. As a result, humans cannot readily interpret the behavious of the trained AI model. It is, therefore, a "black box" in many respects. Consequently, applying AI methods requires a focus on problem-solving, not explaining the correlations and causes that justify a specific decision. In recruiting, however, especially in applicant selection, decisions must not only be made understandably but must also be justifiable in a legally compliant manner. In recent years, there have been new research approaches to "explainable and interpretable deep learning" (see, among others, [5–7]). These approaches aim to make machine-learning-based systems' decision behavior understandable for humans. The rationale may be dispensable in some areas, such as the targeted addressing of applicant groups through performance marketing. When AI increases the reach of relevant applicant target groups, explaining reasons may be dispensable. But only if the re-shaping of the target group through the optimization does not introduce or enhance systematic undesirable bias.

9.2.2 Data Centricity

AI systems such as neural networks react to a specific "stimulus" represented by input data with a "response" suitable for solving the problem, manifesting itself in output data. In this context, the first prerequisite is that it must be possible to describe or capture a decision problem appropriately using data. This data-centricity is often the case in recruiting since the current decision-making process—at least for "hard skills"—has shifted towards assessing applicant characteristics reflected in data. Moreover, AI-based decision support allows for considering data that previously could not be systematically evaluated by humans with reasonable effort due to data complexity or format. On the other hand, it is challenging to support decision-making situations in recruiting with AI in cases where "soft skills" are essential. Soft skills are often not directly identifiable from applicant data or cannot be automatically verified reliably. In practice, these skills only become apparent in an interpersonal interaction (e.g., an applicant interview). Given the challenges mentioned above, it is crucial to identify those areas in recruiting that AI can suitably support.

9.2.3 Data Availability

Even when data can suitably describe the decision-making situations in recruiting, one must pay attention to the availability of corresponding data collections or "data sets" required for training the AI models. "Supervised" training of machine learning methods requires, for example, that data is marked so that certain stimuli in the input are recorded together with defined response or output. These data sets must thus be "labeled" by human experts or automatic processing.

For successful training—a good prediction accuracy, generalisability, and robustness— of any machine learning classification models, a large amount of training data must be available. The necessary data preparation can be very costly. However, reliable sources of such labeled data have emerged over the last decade due to the digitization of HR, e.g., in online job boards, digital HR management systems, etc. Structured data relevant to recruiting can be found, for example, in the ESCO database (European Skills, Competences, Qualifications, and Occupations [8]). ESCO provides a multilingual European classification for skills, competencies, qualifications, and occupations relevant to the labor market and education and training in the EU. It currently contains 3,000 occupations and 13,500 competence concepts. ESCO is an example of an expert-based information system, which can be used to train and verify machine learning systems in the HR context.

On the other hand, there are also unsupervised learning methods that can be used to extract patterns and correlations from unlabeled data. In the coming years, unsupervised learning methods—such as clustering, anomaly detection, curiosity-driven learning, etc.—will likely find their way into recruiting and HR systems. However, these methods require vast amounts of data for acceptable accuracy, which has not yet been collected in the relatively new field of digitized recruiting. Furthermore, working with applicant data is even more difficult, as it usually contains personal and sensitive information (health data, religious affiliation, etc.) that require special protection according to the European General Data Protection Regulation (GDPR) [9]. Additionally, in the recruiting process, data may sometimes only be stored for a limited period and is often fragmented across multiple IT systems. Therefore, it must be noted that, although good use cases can often be identified "theoretically" for the application of AI in recruiting processes, the feasibility in practice may fail due to a lack of suitable training data.

9.2.4   Decision Transparency and Trustworthiness

Decision support utilizing AI requires training machine learning solutions, which ultimately leads to the fact that "behavior patterns" applied in the past are learned and used to solve future problems. However, some patterns would not be desirable to continue (e.g., discriminatory or outdated ones). On the other hand, the deep learning methods and tools are sometimes not sufficiently transparent. As a result, the underlying patterns governing the decision recommendation are not always clear. Additionally, suppose the robustness against special attacks (e.g. data injections and prompt injections) is not explicitly reinforced. In that case, the AI systems are unreliable: trained models can be fooled by adversarial or unusual input. They can lead to inappropriate mispredictions, unethical applicant selection, or evaluations of involuntary personal messages in interviews.

This tendency can harm applicants whose characteristics are underrepresented in historical data. Furthermore, the models may not automatically consider ethical principles (equality, diversity, etc.). All these aspects may prevent their wide acceptance [10]. If, for example, certain applicant groups have been disadvantaged in a company selection, an AI system that is trained based on historical data in the company will learn and perpetuate this behavior. In practice, there are already examples of companies that have rejected corresponding AI solutions for this reason. For example, Amazon tried to develop an AI solution for presorting job applications back in 2014 [11]. However, it was found that this solution led to discrimination against female applicants. This is because the company had primarily hired male applicants in the past. The system learned and adopted this behavior via the data used as training data [12].

An essential step toward solving this problem was taken in 2019. On behalf of the European Commission, the independent "High-Level Expert Group on AI" published the "Ethical Guidelines for Trustworthy AI"; it defined the key requirements for the trustworthiness of AI systems [13]. It should be noted that the evaluation of trustworthiness for concrete use cases requires significant adaptation of existing software evaluation methods to the specific needs of AI. Testing the adaptive behavior of (statistical) learning systems is particularly difficult due to the stochastic nature of their results. Therefore, the problem of the "ethics of AI" must be kept in mind, but by no means should it be considered unsolvable. There are even current approaches to using AI methods to overcome the hidden, involuntary bias in human decisions. See Sect. 9.5 and the study "Semantics Derived Automatically from Language Corpora Contain Human-like Moral Choices" [14], in which it was shown that AI algorithms could learn a "moral compass."

9.2.5 Continuity and Robustness

In recruiting, problematic situations are not only those in which historical data depict decision-making behavior that exhibits deficiencies or distortions. It should also be noted that corresponding systems or trained models do not adjust independently to changes in the applicant situation if there is no continuous feedback and dynamic adjustment of the models. For example, consider the global economic crisis caused by the spread of Covid-19 disease in 2020. Such a singular event would not have been foreseen by any model trained on pre-2020 data, so the models would not describe the job market dynamics in the subsequent years.

For example, the BetterAds feature of milch & zucker (described in Sect. 9.5.1) was trained in January 2020 on the data from the job portal JobStairs [15] to predict the click-through rates and visitor numbers of various job ads. However, visitor occurrence and behavior have fundamentally changed due to the subsequent Covid-19 crisis, which has required re-training and updating this model. Moreover, even if models are regularly adjusted, AI systems cannot respond reliably to discontinuous developments or disruptive events. Typically such events manifest themselves outside the captured stimuli or are not present "systematically" in the training data (e.g., the collapse of applications and necessary reorientation of applicant targeting due to skills shortages). Therefore, an important current research direction is the search for methods that automatically detect whether the use case parameters are significantly outside the range "seen" in the training set.

Overall, it should be noted that there are high requirements for using AI in recruiting, particularly from a technical and organizational perspective. Initially, AI solutions will primarily be used to support individual tasks in recruiting processes. The complete automation of process sections or even largely autonomous, AI-supported decision-making processes is not yet expected in the medium term, even from a technical perspective. In recruiting, as in many socially relevant areas, a joint effort of human and artificial intelligence is required; "augmented intelligence" combines both aspects and improves human decision-making capabilities [16].

9.3   Limitations for the Use of AI in the Recruiting Process

9.3.1   Acceptance of AI in the Recruiting Process

"Users" and "affected parties" are increasingly placing their specific demands on AI in recruiting. By users, we mean not only recruiters, who work with AI-based systems on the company side, but also all other people involved in the process. Although numerous user surveys indicate a fundamental acceptance of AI-based applications in recruiting, they also point to clear limitations that become apparent whenever final staffing decisions are made by a "machine" and no longer by humans. But a support function in the sense of "assistance" by AI in the recruiting process is welcome.

Those affected are, first and foremost, the applicants. The acceptance of AI in recruiting is more significant the closer the use of AI is to the beginning of the candidate journey. For example, an optimization in the playout or design of an online job ad is viewed positively as a "convenience" from the recruiters' perspective. But at the same time, there is the requirement that this task must be done in a non-discriminatory manner. On the contrary, in the case of the final selection tools, the motto "rather a recruiter than an algorithm" still applies in most cases. In general, the transparency requirement is essential across all process steps (see [17], and the literature cited there). On the other hand, the acceptance of AI in recruiting is also likely to depend on whether the employees in HR departments view such solutions as support or reject them because of concerns about being made redundant by AI-based process automation. Studies already exist in chatbots that could prove relationships between automation anxiety and the perceived usefulness of recruiting chatbots [18].

9.3.2   AI and Compliance with Laws and Standards

The use of AI—also in recruiting—is being discussed in politics, legal associations, and science. For example, the German Bundestag has set up an Enquete Commission "Künstliche Intelligenz" (Artificial Intelligence) to review and prepare legislative measures relating to AI and has explicitly focused on AI in recruiting in Project Group 4 "KI und Arbeit, Bildung, Forschung" (AI and Work, Education, Research). The focus is on questions regarding the effects of AI on (employee) data protection (DSGVO, BGSG), anti-discrimination law (AGG), and corporate co-determination (BetrVG).

Professional associations such as the "Bundesverband der Personalmanager" (Federal Association of Personnel Managers) have supported the establishment of an "Ethikbeirat HR-Tech" (HR-Tech Ethics Advisory Board), which has now presented the elaborated "Guidelines for the Ethical Use of AI Technology in HR" [19]. Furthermore, NGOs such as "Algorithm Watch" are commissioning expert reports on "Existing and future regulations on the use of algorithms in HR" [20], and the Federal Anti-Discrimination Agency initiated an investigation into the "Discrimination risks through the use of algorithms" [21].

The scientific community has formulated specific requirements to use AI in personnel selection, particularly regarding job-related aptitude diagnostics. In addition to the general consideration of DIN standard 33430, they propose an evaluation of every AI-based diagnostic tool for compliance with scientific standards using the "Testbeurteilungssystem des Diagnostik- und Testkuratoriums (TBS-DTK)" (Test Assessment System of the Diagnostic and Test Board), for instance, "speech evaluation systems" for the personality trait detection [22]. We refer to [23] for details on the lines of argumentation about what "AI for HR" can and should generally achieve.

9.4   Systematization of Relevant Application Areas of AI in the Recruiting Process

On the one hand, the recruiting process follows the candidate journey. Still, on the other hand, the company's internal organizational units—here, first and foremost, the recruiting department—are directly involved in the various sub-processes of recruiting: from job approval to job posting and selection of posting channels to receipt of applications and applicant selection. Therefore, communication and decision processes must take place in each of the various sub-processes. This applies to communication with external and internal applicants and with the different internal recruiting stakeholders. Above all with the managers who need to fill the vacancies.

For the recruiting organization as "process owner," the question arises, in which of the sub-processes above the use of AI can offer added value. This involves weighing up what is technologically feasible with an expected improvement in the result of the respective sub-process. Last but not least, there is the question of acceptance and whether corresponding AI-based applications in recruiting are allowed by law. Figure 9.1 shows starting points for individual process steps with AI in recruiting.

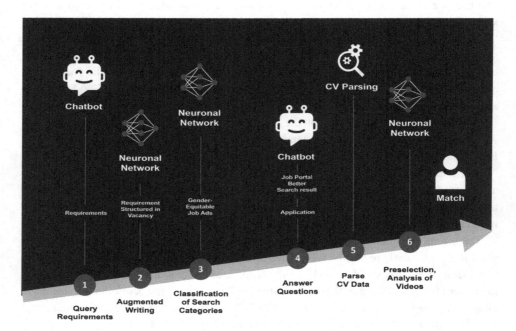

Fig. 9.1 Selected application areas of AI in recruiting

The chatbots shown in Fig. 9.1 are not necessarily linked to using AI for decision-making. Instead, they represent more of a user interface to enable a natural language and automated dialogue with applicants in the recruiting process [24]. However, there are narrow limits to a "real" dialogue between humans and computers without AI or with purely rule-based chatbot solutions based on matching keywords. AI is, therefore, a fundamental prerequisite for chatbots for a "natural" dialogue process and for the fact that free text input from users is understood by the chatbot or an underlying user intention can be extracted based on speech analysis.

In the following, some selected use cases of AI will be described more concretely along the recruiting process. After that, two pilot projects from the German market, where attempts are made to implement corresponding scenarios in practice, are presented in the following chapter.

9.4.1   Formulation of Job Advertisements

The recruiting process regularly begins when a vacancy arises. A vacancy triggers communication between the specialist department ("hiring manager") and the recruiting department or recruiter. Already here (see No. 1, Fig. 9.1), an "intelligent" chatbot can help. For example, a chatbot can query the requirements in the specialist department in a dialog and pass the data on to the HR department in a structured manner.

AI can also assist in creating the job advertisement itself, e.g., AI can help formulate the advertisement texts in gender-appropriate language. This can be done through augmented intelligence, i.e., as support for humans. For example, an augmented writing system (see No. 2, Fig. 9.1) suggests texts or better options for the wording of the job advertisement to the user, who ultimately decides whether to adopt the AI suggestions.

The AI-based program analyzes the ad text when jobs are advertised and determines the national language, entry-level, function, industry, and other characteristics. Based on a data-driven and scientifically sound text analysis, AI can even predict the likely success of a job ad.

9.4.2   Publication of Job Advertisements on Job Boards and Social Networks

Job advertisements must be classified to fit into the media's search categories. For example, for which location, career level, industry, and functional area does the job ad apply? AI can automatically assign the job ad to these categories and criteria via text analysis (see No. 3, Fig. 9.1). Automatic categorization is convenient for companies that use job boards. The classification into the categories of the respective job board is done accurately by AI. The company has no longer to deal with each job board's different categories and criteria. AI support simplifies job ad placement and can help optimize search results.

9.4.3 Structuring and Analysis of Application Data

In the context of applicant management, AI can initially support the process of importing applicant data in the applicant tracking system. Chatbots can be used, for example, to answer an applicant's questions about the application process. Such a chatbot can also support the completion of the questionnaire by asking questions in natural language that the applicant answers (see No. 4, Fig. 9.1). The software then structures the data automatically.

Another form of AI utilization is not dialogue support but is based on its use to analyze the applicant documents uploaded to an applicant management system. A parsing of CV data (see No. 5, Fig. 9.1) can, for example, support the applicant's data entry or relieve them of a time-consuming transfer of such data into input masks. For this purpose, the CV is automatically read from uploaded documents and certificates to pre-fill the applicant form with the extracted data. The applicant can focus on checking the entries for correctness and complete missing data. As a result, the recruiting organization receives structured, high-quality data.

Moreover, AI can support the pre-selection of applications or a corresponding matching of applicant data and job requirements (see No. 6, Fig. 9.1). To relieve recruiters, AI can quickly evaluate and show the fit to the requirements of a job in a structured manner as part of a systematic document analysis. This automized pre-selection can make it easier for a recruiter to identify suitable candidates. In addition, it relieves them of the time-consuming task of "searching out" relevant information from the applicant documents.

AI-based job application selection, on the other hand, is to be distinguished from such support services provided by AI for preparing job application documents. As early as 1917, the problem of assessing the job suitability of candidates was recognized as "The Supreme Problem" [25]. Since then, this applicant matching problem has interested researchers and practitioners [26]. At the same time, caution is required as algorithms can be prone to bias in data and data labels. As noted in Sect. 9.2.4, bias can arise when training AI with historical application data, which then calls into question the objectivity required of such AI-assisted decision recommendations. Conversely, AI can be used to identify or avoid such a "bias." Furthermore, algorithmic selection methods can lead to more diversity, which is also the subject of one of the practical examples in Sect. 9.5.

9.5 Selected Practical Examples of AI in Recruiting

9.5.1 "BetterAds"—Performance Analysis and Augmented Writing of Job Advertisements

The "BeeSite BetterAds" tool is an AI application for performance analysis and augmented writing. The tool is an additional function in the applicant tracking system BeeSite, which is developed and distributed by the German company milch & zucker AG. The product, released in April 2020, helps recruiters improve the formulation of job advertisements through an AI analysis of the text and suggested text improvements ("augmented writing"). The tool is intended to help optimize job ads and thus achieve a higher number of good-fit and diverse applications.

BetterAds analyses the text (title and full text) of a job ad to predict the potential success relative to similar ads on the market. The main metrics of success evaluated here are the number of applications per unique visitor, the number of visitors to a job ad per day (reach), and the applicant diversity. In addition, the software considers the impact of specific wording and keywords on the success of job ads. Finally, it provides an overall score for each ad. Text optimization is made possible through the application of Natural Language Processing [27]. The AI-assisted tool analyzes the text of the job ad and generates suggestions on how to improve the success of the job ad and make the language usage gender-neutral. It also proposes keywords that can be used effectively for a web advertising campaign.

Components of the Job-Ad Analysis
The recruiter interface provides a structured dashboard in which the key analysis results are shown. Some indicators are illustrated graphically. An example of such a performance analysis is shown in Fig. 9.2 and comprises the following components:

- Total score: This score represents the prediction of the success of a job ad. The higher the score, the higher the probability that candidates will find the ad, read it and apply. The total score combines the results for the title score ("clickability"), the length score, and the gender sentiment score, which are explained separately below.
- Title score: A machine-learning model predicts the expected outreach for the ad, i.e., how often it is expected to be viewed per day. The reach prediction is based on analyzing the click rates of current ads. It reflects both the likelihood for the job ad webpage to be

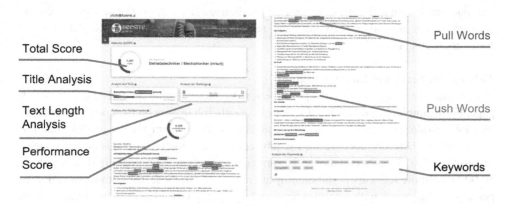

Fig. 9.2 Dashboard of the BetterAd tool with analysis components

found through the web search engines and the attractiveness of the title to the potential applicants, i.e., the probability of being clicked to read the entire ad. The model is regularly re-trained and updated as needed, e.g., after a crisis event such as Covid-19.

- Text length analysis: The length of the ad text compared to similar ads gives recruiters an orientation as to which length is usual in this environment. The length comparison is based on a large pool of up to 2,000 similar ads, individually identified for each ad through an AI-supported text-similarity comparison.

- Gender-sentiment: The score calculation is based on the ratio of scientifically determined "push" and "pull" words and expressions, which also can be viewed as stereotypically associated with a male or female role. Previous research has shown that using fewer push words and more pull words can encourage job seekers to apply more often, especially women. The push and pull words (for the German language) are determined from the psychology studies on gender stereotypes in the workplace and the wording of job advertisements (see [28] and the literature therein). A higher score suggests a higher expected number of applications overall and an increased proportion of women in the applicants. An augmented writing component of the tool provides suggestions for neutral re-formulations of phrases involving push words. Further, the tool identifies the underlying professional title in the job-ad title to assist in the gender-neutral formulations of the titles. Finally, it provides the gender alternative form or gender-neutral variants if available. The recommendation was trained on data from the European ESCO database [8].

- Keywords: Terms characteristic of the ad are displayed which can be used for an online advertising campaign (on Google, Facebook or Instagram) for example. The words are listed according to relevance. For the relevance rating a formula is used that also incorporates information from web tracking of search terms

The selection of the implemented functions results from the requirements described in Sects. 9.2 and 9.3. Large amounts of data were processed to train the models behind the BetterAds tool. Such training was possible using a high-performance computer and modern machine learning libraries that can be parallelized [29–32]. The data came from the company's job posting archive and the real-time webpage-performance data on the JobStairs portal [15].

Title Score: Technical Implementation
The job-ad title is significantly responsible for the success of an advertisement as a web page. For example, in the results of a Google search, job teasers on the company website, or a job newsletter, there is often very little information except for the job-ad title. The job-ad title is therefore decisive for the job seeker's decision to open this advertisement, i.e., to "visit" it. To collect data for the outreach prediction, web-tracking data for about 50,000 ads on the Jobstairs portal monthy are recorded. For these, the number of activities per day is considered, with activities representing mainly visits, but also, to a lesser extent, clicks and downloads. The ad activity index is defined in a range from 0 to 1, with 0.5 being the value for the average ad.

The statistical analysis has shown that the distribution of ads is not symmetrical. Instead, an "influence effect" can be seen. An index close to 1 is achieved for less than 5 percent of ads. These do not get twice as many visitors as the average ads (index = 0.5) but up to 40 times more. This is usually the result of potentiation by Google. Someone visits the website, and due to this, Google classifies it as more relevant; thus, it appears more often in the search result, and more people read the ad, etc.

A custom deep neural network learns to predict the activity index for new (unknown) ads through a non-linear regression based on the job ad titles on this collection of ads with known indices. The prediction accuracy is over 90 percent on average for the current ads. So the program can analyze whether any title will likely lead to many or few visitors to the job ad. It must be mentioned that the trained model does not remain applicable after a significant change in the applicant situation. For example, the emergence and behavior of job seekers will most likely change due to the current Covid-19 crisis. Therefore, it is necessary that the neural network is continuously trained and the corresponding tool function is updated (see Sect. 9.2.5).

Further Development of the BetterAds Tool
Possible future development for the BetterAds tool could focus on the interactive, semi-automatic generation of complete texts for job advertisements. In recent years, variants of Generative Adversarial Networks (GANs) seemed as a promising tool for this purpose in academic research [33, 34]. GANs perform very well in image generation [35] but have proven challenging to train in the natural language domain. Most recent innovations, such as the successful "Transformer" models based on attention mechanisms, can be used instead [36]. Automatic text generation is a springboard for a wide range of applications, from job ad generation and machine summarization of job applications to dialogue generation, which becomes relevant for the following section on chatbots.

9.5.2   "CATS"—Recruiting Chatbots in Applicant Tracking Systems

In recent years, the recruiting channels of companies have adapted to changes in media usage behavior. Whereas the focus used to be on the one-way transmission of information from one "sender" (employer) to many "recipients" (applicants), online communication is increasingly changing to a more dialogic conversational approach. The two-way exchange of information is gaining in importance. A corresponding development has taken place in recruiting. Information for applicants is no longer only "published" on career websites and job portals. Companies are trying to implement supplementary social media offerings, for example, to make more direct contact with target groups. Especially among younger target groups, other forms of interaction, such as intensive use of messaging services, have become established in recent years in addition to social media. Moreover, mobile devices such as smartphones and tablets are becoming increasingly crucial for job searches and applications (see [37] for more details).

The traditional "search and find" of information on web pages and the retrieval of information via desktop browser is losing importance. New forms of interaction that are more closely related to "questions and answers" emerge. They support a dialogue via text or voice inputs to satisfy job seekers' information needs. Against this background, chatbots are computer-based systems that support a natural language dialogue between humans and machines and can automate dialogue-based communication processes [24].

Functions of Chatbots

As shown in Fig. 9.3, chatbots are typically integrated via chat windows on websites or in messaging applications. The user enters his question as a short text and receives a quick answer, just like in a chat. However, this can contain text and buttons to structure further the dialogue or media content, such as images, or website links. Chatbots can be distinguished from so-called voice assistants such as Alexa, Siri, or Google Assistant, which function similarly but do not process text input but interact with the user via spoken language. Chatbot systems are not new, but their applicability in practice has recently improved significantly due to advances in AI. In contrast to rule-based systems, which, for example, link the course of the dialogue to the identification of specific key terms or keywords in the user input, the use of AI for natural language processing [27] opens up a much more flexible and thus also more robust dialogue guidance.

Fig. 9.3 Example of a chat history in the CATS chatbot

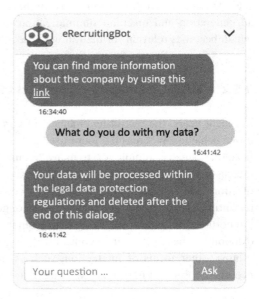

Today, there are numerous platforms for the development of chatbots. However, implementing efficient chatbot systems and their integration into existing recruiting processes is still a complex and challenging task. This is especially true if the introduction of chatbots is not only intended for marketing purposes to convey an innovative employer image but to reduce the workload of recruiters and improve recruiting processes. Therefore, the following aspects should be examined to narrow down suitable areas of application:

- Dialogue reference: Information needs can be better satisfied in a dialogue process.
- Real-time accessibility: It is crucial to answering requests without delay.
- Increased efficiency: Automated dialogues meet demand better or more cost-effectively.
- High case numbers: Many queries justify the development effort and provide training data.

Chatbots in Applicant Support

The project "Chatbots in Applicant Tracking Systems (CATS)" was a research project of the RheinMain University of Applied Sciences in Germany in cooperation with the company milch & zucker AG from 2018 to 2020. The project investigated recruiting chatbot use cases and its potentials, especially in applicant tracking systems [38]. The project aimed to develop a recruiting chatbot toolbox based on available technologies. A prototype of this toolbox was to be integrated into an applicant tracking system. With an easy-to-use graphical user interface, recruiters should be able to select individual chatbot modules and configure them in such a way that they can support applicants in the recruiting process before, during, and after the application.

A concrete example is the chatbot implementation in milch & zucker's applicant tracking system (ATS) "BeeSite." The chatbot service layer in BeeSite will consist of a set of ready-to-use building blocks that operators can easily combine. With graphical user interfaces, models, tools, and data sets provided by milch & zucker AG, the chatbot solutions can be flexibly adapted to the needs of individual companies. The simplified basic structure of the CATS chatbot construction kit is shown in Fig. 9.4 using four main chatbot modules as examples. The chatbot solution is integrated as a functional extension into BeeSite. One of the main components is the frontend, through which users can communicate with the chatbot. In addition, three backend components support the administration of the chatbot modules. In the configuration component, for example, pre-defined sets of questions and answers can be selected and edited. Such sets can include, for example, Frequently Asked Questions (FAQs) about the operation of the ATS or basic FAQs with answers to questions about the company in general. Questions (e.g., "How should I upload my documents?" or "What file format should my documents have?") are categorized by information needs in the form of so-called "user intents" (e.g., "File format for documents"). For each intent, a larger amount of variants of questions with the same intention is required. These questions are used as training data to train the "intent matching," i.e., the AI-based classification of the users' information needs. The

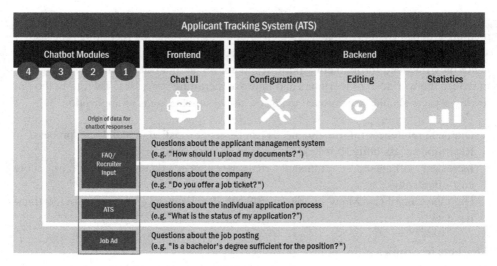

Fig. 9.4 Basic components of the CATS chatbot architecture

pre-defined data sets can be extended and adapted to company-specific requirements via the configuration component.

During the chatbot's operation, users' questions will arise in the system that cannot be assigned to a user intent by the AI algorithms. There may also be questions about content that has not yet been considered as intents. For these intents, no suitable answers can be provided. All such user input is reported in an editing component of the chatbot system. The editing component provides functions to assign reported questions to user intents manually. These questions will then be used as training data to improve the intent matching. The editorial component can also be used to enter new intents and related training questions and answers.

Moreover, intents and associated responses could be retrieved by integrating information from the ATS or analyzing the text of job ads. The implementation of these chatbot modules 3 and 4 in Fig. 9.4 is significantly more complex. First, contextual information must be obtained to operate these modules properly ("Which specific applicant is interacting with the chatbot?" or "For which job ad is the application?"). Second, interfaces are required to request data from the ATS (e.g., the status of an application), and an advanced knowledge base must be available to interpret job ad information (e.g., if a specific education is sufficient to apply). However, it is precisely the answering of such questions that can generate considerable added value for applicants. Especially if answers are provided quickly, or without delay, as with conventional inquiries, significant potential can be expected here regarding relieving recruiters. Again, however, the technical implementation represents a considerable challenge.

In the CATS project, an initial solution was prototypically implemented, which can be assigned to chatbot modules 1 and 2 in Fig. 9.4. The CATS chatbot is concerned with answering questions about the application process in the applicant management system and the company. For the technical implementation, the platform Rasa was chosen [39]. Rasa is a machine learning-based open-source framework that provides various components for a Natural Language Understanding (Rasa NLU) and a dialogue engine (Rasa Core) to develop chatbot solutions or voice assistants. In recruiting, using an open-source framework has the advantage that it can be easily operated "on-premise," i.e., on the company's servers. This is particularly important if the chatbot is used with personal data from applications. In the CATS project, the developed chatbot solution was trained with application data or inquiries received by the applicant support of a large German DAX company. Since this company does not yet use a chat or chatbot, the user intents were extracted from mail requests. Additional question variants were generated for a fictitious chatbot-based user request. With the model trained in this way, theoretically, about 70 percent of the support requests received by mail could be answered by a chatbot in the future. However, the extent to which the solution is suitable for everyday use and is accepted by applicants as an alternative to mail queries still has to be tested or can only be proven once the prototype has been integrated into the applicant management system in practice.

9.6   Summary and Outlook

Many opportunities exist to use AI to support recruiting processes. AI is already used for the optimized creation and publication of job ads. Chatbots are used to automize the answering of standard requests from applicants. Pilot projects have achieved promising results but are still at the beginning of their development. However, all these AI applications in recruiting are "supporting AI solutions" and far from being able to replace human recruiters. The current status quo of recruiting AI is primarily suitable for relieving recruiters of frequently occurring routine processes. Limits to using AI are not only technically based on the development status of AI. Successful implementation of machine learning requires suitable training data in high quality and huge quantities. In addition, AI applications in recruiting must also be designed so that the results are acceptable and understandable for all stakeholders involved in the recruiting process. Furthermore, legal regulations and requirements in the recruiting area must be fulfilled.

AI solutions have constantly been evolving through the ups and downs of euphoric technology forecasts and disappointed expectations for several decades. Finally, however, there are indications that a level of performance has been reached today at which AI-based solutions can efficiently support business processes in recruiting. When the development and acceptance of AI continue to progress at the current rapid pace, AI will no longer mainly serve as marketing and sales arguments but should soon be a common and indispensable productive tool in recruiting processes.

Funding Notice

The project "CATS—Chatbots in Applicant Tracking Systems" (HA-Project-No.: 642/18–6 5) is funded within the framework of Hessen ModellProjekte by the "LOEWE— Landes-Offensive zur Entwicklung Wissenschaftlich-ökonomischer Exzellenz, Förderlinie 3: KMU-Verbundvorhaben".

References

1. Malin L, Jansen A, Seyda S et al (2019) KOFA-STUDIE 2/2019, Fachkräfteengpässe in Unternehmen. Fachkräftesicherung in Deutschland—diese Potenziale gibt es noch. https://www.kofa.de/fileadmin/Dateiliste/Publikationen/Studien/Fachkraefteengpaesse_2019_2.pdf. Accessed: 4. Mai 2020
2. Anger C, Koppel O, Plünnecke A et al (2019) MINT-Frühjahrsreport 2019: MINT und Innovationen—Erfolge und Handlungsbedarfe. https://www.iwkoeln.de/fileadmin/user_upload/Studien/Gutachten/PDF/2019/MINT-Fr%C3%BChjahrsreport_2019.pdf. Accessed: 4. Mai 2020
3. Bitkom/DFKI (2017) Künstliche Intelligenz: Wirtschaftliche Bedeutung, gesellschaftliche Herausforderungen, menschliche Verantwortung. https://www.dfki.de/fileadmin/user_upload/import/9744_171012-KI-Gipfelpapier-online.pdf. Accessed: 4. Mai 2020
4. Goodfellow I, Bengio Y, Courville A (2016) Deep learning (adaptive computation and machine learning). MIT Press, Cambridge
5. Escalante HJ, Guyon I, Escalera S, Jacques J, Madadi M, Baro X, Ayache S, Viegas E, Gucluturk Y, Guclu U, van Gerven MAJ, van Lier R (2017) Design of an explainable machine learning challenge for video interviews. In: 2017 international joint conference on neural networks (IJCNN). IEEE, p 3688–3695. https://doi.org/10.1109/IJCNN.2017.7966320
6. Liem CCS, Langer M, Demetriou A, Hiemstra AMF, Sukma Wicaksana A, Born MP, König CJ (2018) Psychology meets machine learning: interdisciplinary perspectives on algorithmic job candidate screening. In: Escalante HJ, Escalera S, Guyon I, Baró X, Güçlütürk Y, Güçlü U, van Gerven M (Hrsg) Explainable and interpretable models in computer vision and machine learning. Cham, Springer, p 197–253. https://doi.org/10.1007/978-3-319-98131-4
7. Du M, Liu N, Hu X (2019) Techniques for interpretable machine learning. Commun ACM 63(1):68–77. https://doi.org/10.1145/3359786
8. ESCO (2020) Database for European skills, competencies, qualifications, and occupations. https://ec.europa.eu/esco/portal/home. Accessed: 4. Mai 2020
9. DSGVO (2018) Verordnung (EU) 2016/679 (Datenschutz-Grundverordnung) in der aktuellen Version des ABl. L 119, 04.05.2016; ber. ABl. L 127, 23.05.2018. https://eur-lex.europa.eu/legal-content/DE/TXT/PDF/?uri=CELEX:32016R0679. Accessed: 4. Mai 2020
10. Mayson SG (2019) Bias in, bias out. Yale Law J 128(8):2218–2300
11. Dobson S (2018) Is AI biased in recruitment? Amazon's misstep highlights challenges for employers. https://www.hrreporter.com/focus-areas/hr-technology/is-ai-biased-in-recruitment/299161. Accessed: 4. Mai 2020
12. Wilke F (2018) Künstliche Intelligenz diskriminiert (noch). https://www.zeit.de/arbeit/2018-10/bewerbungsroboter-kuenstliche-intelligenz-amazon-frauen-diskriminierung. Accessed: 2. Mai 2020
13. EU (2019) Eine Definition der KI: Wichtigste Fähigkeiten und Wissenschaftsgebiete. https://ec.europa.eu/newsroom/dae/document.cfm?doc_id=60664. Accessed: 4. Mai 2020
14. Jentzsch S, Schramowski P, Rothkopf C, Kersting K (2019) Semantics derived automatically from language corpora contain human-like moral choices. In: Conitzer V, Hadfield G, Vallor

S (Hrsg.) Proceedings of the 2019 AAAI/ACM conference on AI, ethics, and society. ACM, New York, NY, USA, p 37–44. https://doi.org/10.1145/3306618.3314267

15. Jobstairs (2020) Bewerber mit Erfolg erreichen, begeistern und binden. https://www.jobstairs.de/de/informationen/fuer-unternehmen. Accessed: 4. Mai 2020

16. Kirste M (2019) Augmented Intelligence—Wie Menschen mit KI zusammen arbeiten. In: Wittpahl V (Hrsg) Künstliche Intelligenz. Springer, Berlin, p 58–71. https://doi.org/10.1007/978-3-662-58042-4

17. Jäger W, Teetz I (2019) Der ganz große Ansatz ist noch nicht in Sicht—Status quo und Entwicklungen von KI im Recruiting. Personalführung 52(6):16–22

18. Eißer J, Torrini M, Böhm S (2020) Automation anxiety as a barrier to workplace automation: an empirical analysis of the example of recruiting chatbots in Germany. In: Proceedings of the 2020 computers and people research conference (SIGMISCPR). ACM, New York. https://doi.org/10.1145/3378539.3393866

19. Straub R (2020) Richtlinien für den ethischen Einsatz von KI-Technologie in HR. Personalmagazin 22(4):54–57

20. Lewinski K, de Barros Fritz R, Biermeier K (2019) Bestehende und künftige Regelungen des Einsatzes von Algorithmen im HR-Bereich. https://algorithmwatch.org/wp-content/uploads/2019/10/Gutachten-Algorithmen-im-HR-Bereich-von-Lewinski-2019.pdf. Accessed: 4. Mai 2020

21. Orwat C (2020) Diskriminierungsrisiken durch Verwendung von Algorithmen. https://www.antidiskriminierungsstelle.de/SharedDocs/Downloads/DE/publikationen/Expertisen/Studie_Diskriminierungsrisiken_durch_Verwendung_von_Algorithmen.pdf?__blob=publicationFile&v=4. Accessed: 4. Mai 2020

22. Kersting M (2019) Wovon man nicht reden kann, darüber muss man schweigen. Reflexionen zur Eignungsdiagnostik mit Precire Personalmagazin 21(11):62–65

23. Strohmeier S (2020) Künstliche Intelligenz in HR—Eine Gefahr? Personalmagazin 22(3):38–39

24. Böhm S, Meurer S (2018) Potenziale mobiler Technologien für HR (Mobile HR). In: Petry T, Jäger W (Hrsg) Digital HR. Haufe Group, Freiburg, p 137–168

25. Hall GS (1917) Practical relations between psychology and the war. J Appl Psychol 1(1):9–16. https://doi.org/10.1037/h0070238

26. Ployhart RE, Schmitt N, Tippins NT (2017) Solving the supreme problem: 100 years of selection and recruitment at the journal of applied psychology. J Appl Psychol 102(3):291–304. https://doi.org/10.1037/apl0000081

27. Hirschberg J, Manning CD (2015) Advances in natural language processing. Science 349(6245):261–266. https://doi.org/10.1126/science.aaa8685

28. Böhm S, Linnyk O, Kohl J, Weber T, Teetz I, Bandurka K, Kersting M (2020) Analysing gender bias in IT job postings: a pre-study based on samples from the German Job Market. In: Proceedings of the 2020 computers and people research conference (SIGMISCPR). ACM, New York. https://doi.org/10.1145/3378539.3393862

29. Abadi M, Agarwal A, Barham P et al (2016) TensorFlow: large-scale machine learning on heterogeneous distributed systems. https://arxiv.org/pdf/1603.04467v2

30. Bojanowski P, Grave E, Joulin A et al (2016) Enriching word vectors with subword information. Trans Assoc Comput Linguist 5:135

31. Joulin A, Grave E, Bojanowski P et al (2016) Bag of tricks for efficient text classification. https://arxiv.org/pdf/1607.01759v3

32. Spacy (2020) Industrial-strength natural language processing. https://spacy.io/. Accessed: 4. Mai 2020

33. Fedus W, Goodfellow I, Dai AM (2018) MaskGAN: better text generation via filling in the _____. https://arxiv.org/pdf/1801.07736v3

34. Yu L, Zhang W, Wang J et al (2016) SeqGAN: sequence generative adversarial nets with policy gradient. Proc AAI Conf Artif Intell 31(1)
35. Goodfellow IJ, Pouget-Abadie J, Mirza M, Xu B, Warde-Farley D, Ozair S, Courville A, Bengio Y (2014) Generative adversarial nets. In: Proceedings of the 27th international conference on neural information processing systems, vol 2. MIT Press, Cambridge, p 2672–2680
36. Vaswani A, Shazeer N, Parmar N et al (2017) Attention is all you need. https://arxiv.org/pdf/1706.03762v5
37. Böhm S, Jäger W (2019) Spurenelemente: Künstliche Intelligenz im Recruiting steckt (noch) in den Anfängen. Personal Manager o.Jg.(6):20–23
38. CATS Chatbots in Applicant Tracking Systems. https://www.hs-rm.de/de/fachbereiche/design-informatik-medien/forschungsprofil/cats. Accessed: 2. Mai 2020
39. Rasa (2020) Build contextual assistants that really help customers. https://rasa.com/. Accessed: 2. Mai 2020

Stephan Böhm is a professor of Telecommunications Technology/Mobile Media in the Media Management program at RheinMain University of Applied Sciences since 2006. After receiving his diploma in industrial engineering from the TU Darmstadt, he completed his doctoral thesis on innovation marketing for UMTS mobile services at the University of Duisburg. Stephan Böhm then worked as a management consultant for Booz Allen Hamilton for several years. He has a longstanding experience in telecommunications and media markets. As an expert in mobile media, Stephan Böhm is a member of various expert panels and committees, speaker at specialist events, and is the author of over 90 publications. He is co-initiator of the Mobile Media Forum in Wiesbaden and the Center of Advanced E-Business Studies (CAEBUS) at the RheinMain University of Applied Sciences.

Olena Linnyk heads the machine learning team at milch & zucker AG and is responsible for Natural Language Processing and pattern recognition from incomplete and noisy data. The latter has also been the focus of her active research and teaching for years. Olena Linnyk lectures on Artificial Intelligence at the Departments of Mathematics and Computer Science, Physics, Geography, and Medicine at the Justus Liebig University Giessen. She conducts research in the "DeepThinkers" group at the Frankfurt Institute for Advanced Studies. Olena Linnyk has contributed to several projects on applying machine learning in basic research and developing new AI-based products and services. She is the author of more than 110 peer-reviewed journal and conference papers.

Wolfgang Jäger was a professor in the Media Management program at RheinMain University of Applied Sciences in Wiesbaden from 1995 to 2018. From 2018 to 2020, he has been in an advisory position in the LOEWE research project CATS. Wolfgang Jäger studied business administration at the University of Essen. In 1984 he received his doctoral degree from the University of the Federal Armed Forces in Hamburg with a topic on co-determination in the workplace. His research focuses on optimizing human resources and communication-related processes, structures, and management systems. He conducts consulting and practical projects on these topics, regularly chairs conferences, and writes numerous articles and books. Since 1990, he has been a partner in Dr. Jäger Management-

Beratung and DJM Consulting GmbH in Königstein im Taunus and has received several awards as a "leading" HR consultant.

Ingolf Teetz is an expert in recruiting processes and software. For over 20 years, Ingolf Teetz has accompanied large and medium-sized companies in implementing recruiting and talent management software with his experience in international HR processes and enthusiasm for programming software. He is a physicist, co-founder of milch & zucker— The Talent Management and Software Company AG, and CEO of the company since 2014. He is responsible for the Recruiting Software and Technology division, whose main product has repeatedly been awarded the title of "Germany's Best Applicant Management System." In addition, Ingolf Teetz is a member of the Board of Directors of the international HR Open Standards Consortium, which develops standards for the exchange of HR data to simplify the integration of HR systems.

Artificial Intelligence in the Staffing Process: Performance Comparisons of (Un)supervised Learning for the Screening of Job Applications

Marc Roedenbeck, Salmai Qari, and Marcel Herold

Abstract

Artificial intelligence has the potential to change not only technology related jobs but also administrative jobs. While work on artificial intelligence has a relatively long tradition in research areas such as computer science or statistics, human resource management is an area in which the topic has only a short history. There are many proposals for the use of artificial intelligence in human resource management, but there is a lack of empirical examples. This paper attempts to exemplify a concrete use case through a pilot study with three small datasets. Using simple cluster analysis from unsupervised learning and a neural network from the supervised learning methods, application letters are analyzed. Depending on the quality and size of the dataset, it can be shown that cluster analysis can significantly replicate human judgment in terms of application quality groups via document-word matrices and furthermore determine the optimal number of groups for judgment. Thus, recruiters would only need to look at a few applications for each determined group in order to determine the quality of the specific group. With the use of the neural network, an increase in performance could be achieved both in the replication of human judgments and in the case of the optimal number groups for judgment depending on the number of hidden neurons used. However, this is limited to the training data set. In this respect, it can be concluded that the use of both

M. Roedenbeck (✉) · M. Herold
Wildau University of Applied Sciences, Wildau, Germany
e-mail: marc.roedenbeck@th-wildau.de; mherold@th-wildau.de

S. Qari
HWR, Berlin, Germany
e-mail: salmai.qari@hwr-berlin.de

T. Barton, C. Müller (eds.), *Artificial intelligence in application*,
https://doi.org/10.1007/978-3-658-43843-2_10

methods represents a promising approach for the introduction of artificial intelligence in recruiting.

10.1   Artificial Intelligence in Human Resources Management

Digitalization and the increase in the use of technology is changing the world of work through rapidly growing scientific developments such as nanotechnology, robotics, machine learning, algorithms and artificial intelligence. Companies are forced to deal with this complex issue in order to gain a competitive advantage. The German federal government has also recognized the need and decided to promote and expand research in the area of Artificial Intelligence (AI) [1]. AI is a digital system that interprets external data, learns from it, and can solve tasks and achieve goals through adaptation [2]. AI is believed to be a driver for societal advancement as there is a great potential benefit for the labor market: While some focus on AI as a driver of growth [3, 4], others see it as a driver of technologically induced unemployment [5].

The use of AI in the context of classification issues e.g. in object recognition, in particular neural networks as one of its more prominent methods, are long established in the fields of mathematics, statistics and computer science (see e.g [6, 7]). Neural networks can learn to uncover non-linear patterns in incomplete datasets. These results can in turn be used for predictions of various kinds (see, e.g., [8–11]), or find their way into valid decision making [12, 13]. In organizational science, the use of neural networks started only in the late 1980s (see, e.g., [14–18]), where they have been used for exchange rate predictions or to detect credit card fraud.

The integration of AI in human resource management is mostly limited to considerations of what specific jobs could look like in the future, taking AI into account. It is also considered which jobs can be taken over by AI due to digitalization. The substitution of jobs has increased from 15% to 25% due to the advancement of technology [19]. However, the degree of substitution varies considerably across sectors and over time [20]. The use of AI in human resource management, on the other hand, has been studied to a very limited extent [21]. HR managers do not yet perceive the potential of this technology for their own department [22]. Yet, the literature already illustrates different approaches for specific tasks, such as in human resource planning using genetic algorithms [23] or in the field of performance management [24]. The non-use of AI is surprising in the sense that a large amount of different personal data about employees is available in human resource management. Few large TECH companies like Google or IBM are already using employee data to predict their future behavior. Google combines different employee data to thereby gain insights for building high performance teams [25]. IBM, on the other hand, uses employee data to make predictions about employee turnover [26]. However, a current study shows, that in the context of personnel management, the application of AI in staffing will play an important role in the future—the greatest application potential is seen in the areas of recruiting and screening of applicants via an automated document analysis [21].

10.2   How Artificial Intelligence Could Be Applied in Recruiting and Screening

Selecting the right candidate for the right position is one of the essential tasks in personnel management, and it is an achievement by a team consisting of HR recruiters, hiring managers of other departments and, if applicable, members of the working council. In small enterprises, it is often done by the top management only. The top row in Fig. 10.1 illustrates the staffing process. The starting point is the identification of job specific knowledge, skills, abilities and other aspects like personal traits (KSAO). The next step is the recruitment which is done either passively (via walk-in applications or unsolicited submissions) or actively (via job postings on various platforms or headhunters and social media search). In the subsequent screening, the applicant pool is adjusted, excluding unsuitable applicants. Therefore, mostly the application documents (cover letter/letter of motivation, resumé/curriculum vitae, references/certificates) are reviewed and evaluated according to predefined criteria. The remaining candidates are then subjected to various tests; the simplest form is an interview, a more complex procedure would be an assessment center. Finally, all assessments are compared and an offer is made for best fitting candidate [27].

This staffing process can be automated with the help of AI, which is highlighted in light grey in Fig. 10.1 (cf. Figure 10.1). Strohmeier and Piazza [23] suggest different possible applications, e.g. the use of knowledge-based search engines in recruiting. Here, the focus is on bridging the gap between the language used by companies and candidates, which are assumed to be different. Thus, there is no correspondence between the terms searched by potential candidates and those used by companies when writing their job postings. An intelligent search engine could close this gap. Another field for the application of AI lays in the screening via semantic analysis based on text mining (tokenizing) . In this process, application documents can be analyzed in relation to the words used, identifying positive and negative ones. A classification algorithm can then be trained with a set of given documents, so that it can evaluate new documents in the future (e.g., motivation letters).

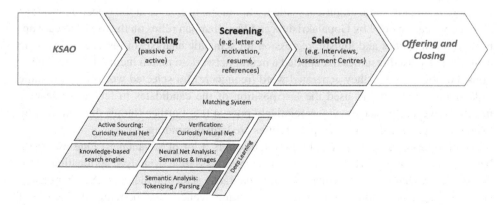

Fig. 10.1 Staffing process modified with and AI perspective

Semantic analysis based on information extraction (parsing) can also be applied to application documents, especially to resumes. Using rule-based algorithms or statistical methods, documents can be evaluated to identify specific information and extract the relationships between entities.

Roedenbeck [24] also shows various possible applications for the use of AI in recruiting: In the recruiting phase of the selection process, HR recruiters sometimes identify candidates through active sourcing. Thereby, they could be supported with a neural network using curiosity learning. A search algorithm based upon a semantic analysis identifies potential candidates via their skills, abilities and competencies presented on social and professional internet platforms (e.g. LinkedIn, Xing, Facebook). The recruiter then decides upon good and bad suggestions and the neural network work in the background creatively adapts and generates new suggestions. In the screening phase of staffing, a neural network could be used as an additional method during the semantic analysis or to analyze image representations of the application documents classifying them into good and bad. A more complex algorithm could be designed relying on a neural network approach with curiosity learning to verify information offered by the candidates on social and professional online platforms. An even more complex approach would be linking all different methods in a deep learning approach. All methods for the individual tasks are interlinked in a superordinate neural network.

Fichtner et al. [22] even go one step further and outline in a theoretical way how the recruitment and screening process could be automated with a matching system. On the one hand, the system could suggests potential candidates to companies and on the other hand, it recommends suitable jobs to candidates. Therefore, the fit between job postings as well as application documents is analyzed. The scheduling of the interview is done by an AI assistant and a machine conducts it. The interview could then be automatically recorded and scored for personality traits, voice, and word choice by the system. After the evaluation of the interviews, the machine decides whether to make an job offer to the applicant or not. While Fichtner also cover processes running after the selection process, these are out of scope of this article.

There is already some empirical evidence about the use of AI in the selection process: A first application done by Gopalakrishna and Varadharajan relies on the use of linear support vector machines and logistic regression for semantic analysis of resumés [28]. Deng et al. [29] also used resumés but applied a neural network to match them with future positions. For this purpose, they generated word frequencies for selected words in the resumé (tokenizing, parsing) and used the last position of the candidates to let the AI learn a match. However, the accuracy was relatively poor at 0.5. Further studies that empirically validate the previously proposed possibilities are largely pending.

The proposed automated matching system offers a lot of advantages: One important benefit would be time savings, as HR recruiters need less time to analyze application documents and thus have more time available for qualitative selection methods. Another benefit is cost savings, as outsourcing to external recruiting agencies can be reduced. The use of AI also increases the quality in recruitment at the same time, as AI can process a high volume of data as part of the screening phase of the selection process [30]. However, the implementation of AI in human resource management also carries a major risk: fear. Due to the high complexity and a possible loss of control, it is necessary that the decisions made are traceable and replicable by the AI so that the results can be trusted and are reliable [31]. The goal of using AI in recruiting should be to support individuals in performing their extensive tasks. Based on the previously explained possibilities for automation and the first empirical evidence found, a high potential can be seen especially in the phase of screening applicant documents. However, since different approaches have been proposed, the following selective question will be investigated in this chapter: How can Artificial Intelligence support the screening phase of the selection process in particular for the analysis of application documents, especially for the analysis of letters of motivation/cover letters? Aspects that are touched by this question are highlighted in dark grey in Fig. 10.1 (cf. Fig. 10.1).

10.3   The Dataset for Method Piloting and Data Preprocessing

The present dataset comprises three partial datasets. The reason for this lies in the data collection, as this took place in the context of three different student cohorts. Twice in the subject "Human Resource Management 1" (PM) as a major in the Master's program "Business & Law" (class of 2018/2019) and once in the subject "Human Resource Management" (HRM) as a minor in the Master's program "Business Management" (class of 2018). Students had to submit an anonymized application (letter of motivation & resumé, name corresponded to four randomly chosen letters) to a fictitious job as one part of a study-accompanying exam. The job posting differed for the two PM cohorts: the cohort in 2018 applied to the position of a Business Lawyer and the cohort in 2019 to a Junior HR Business Partner [HR BP]. The HRM cohort used the second advert [HR BP] only.

Another part of the study-accompanying exam was to evaluate the submitted applications of the fellow students according to a given set of criteria (a guided peer review process). As our research question is focused on the analysis of letters of motivation/cover letters, only the evaluation of those will be presented here: In addition to formal aspects (nine items, e.g. "Is there a salutation?") and linguistic criteria (eight items, e.g. "Is the cover letter worded sympathetically?"), an overall judgment of the cover letter had to be given on a 100% scale in increments of 10 (e.g. 100% "completely convinced", 0% "not at all convinced"). For the present analysis, we only use the overall judgement, not the items on the formal aspects and linguistic criteria; the data set was processed using the statistical software R [32].

Before starting the actual analysis with the selected machine learning methods, data pre-processing was performed in three main steps: The first step required grouping the overall judgments from all student raters per individual application. This was done using the R.dplyr package [33] and the function "group_by". For grouping, the mean and median of the overall judgements were calculated. Descriptive statistics for all three datasets have been generated with the "summary" function from R.base. In the provided overview (cf. Table 10.1) the students' results are obviously widely scattered. In the first partial data set almost the entire scale from 18.75 to 82.22 is used, in the second and third partial data set the assessment is somewhat more restricted between 45 and 86.92 and between 37.33 and 80. Due to the same job advertisement to which the students of set 2 and set 3 had to apply for [HR BP], a combined data set could be created (Comb.). The dispersion within this set is similar to the results of the subsets between 39 and 94.62. All mean values for the sets are between 60 and 70. This suggests that there is a distribution shifted towards the positive judgment in all sets; which is one of the many possible risks in the so-called peer evaluation [34, 35]. A final look at the N of the respective sets clearly shows that they are relatively small and that the application of supervised learning methods is very risky. Nevertheless, due to the idea of a methodological pilot study, we want to take that risk because the size of the datasets corresponds to those of small and medium-sized enterprises with a poor applicant situation and at the same time few or no independent resources for recruiting. In this respect, the analysis here can be regarded as a pilot for special cases.

The second step in the data preprocessing was the preparation of the application letters: For this purpose, the applications (PDF format, stored in three directories) were first read in as text corpus using the R.tm package [36, 37] and the function combination "Corpus/readPDF", respectively. Then, the corpus could be transformed into a standardized form using various transformation functions of the R.tm package: These include transformations in lowercase (tm_map: toLower), without numbers (tm_map: removeNumbers), without punctuations (tm_map: removePunctuation), without spaces (tm_map: removeWhitespaces), and without filler words (tm_map: removeWords[stopwords "german"]). Subsequently, the remaining words in the standardized corpi were stemmed to their word core. This is done with a Stemmer algorithm (Porter) from the package R.SnowballC and the function "wordStem [str, language = "german"]", whereas the alternative packages

Table 10.1 Overview of data stock—overall assessment of cover letter

	Set 1	Set 2	Set 3	Comb. Set
Vintage	2018	2019	2018	(mixed)
Module evaluation	PM1, deepening	PM1, deepening	HRM, minor subject	(mixed)
Site	Business lawyer	HR BP	HR BP	HR BP
Min	18.75	45	37.33	39
1st quartile	53.28	62.86	54.68	60.33
Median	67.5	70	62.92	66
Mean value	62.4	70.43	61.29	66.75
3rd quartile	74.14	79.29	69.85	72.11
Max	82.22	86.92	80	94.62
N[a]	24	23	18	39

[a]The number N refers to the set of grouped judgments, i.e. the applications submitted; but not to the total number of judgments available

R.Hunspell and R.Unive were not verified in detail. Finally, the standardized and stemmed corpi could be converted into one document word matrix (DTM) using the function "DocumentTermMatrix" from the R.tm package. In this matrix, per document (rows) the identified words (column) are counted and their frequency is stored (cell).

The third step then involves merging the grouped judgments from step 1 with the DTM from step 2. This could be achieved for each dataset, including the combined dataset, using the "merge" function from R.base, so that the only the median of the raters for the overall judgment was stored in each document row.

10.4 The Methods Applied of (Un)supervised Learning and Research Hypotheses

Machine learning is considered a subarea of artificial intelligence. The variety of methods in the context of "machine learning" is immense [38], but an essential subdivision feature is the question of unsupervised or supervised learning. This differentiation essentially means that either a certain variable (response variable) is to be predicted with other variables (supervised learning), or that structures are to be recognized without the presence of a response variable (unsupervised learning) [39].

Cluster analysis (Ward.D, K-means), for example, belongs to the area of unsupervised learning and determines the affiliation to clusters on the basis of the Euclidean distances between individual objects. Methods from the area of supervised learning are, for example, classification or regression methods such as multiple linear regression, kNN (k nearest neighbors) or neural networks (e.g. feedforward and/or multiple backward propagation). All of the methods mentioned above require a response variable (e.g. the aforementioned overall judgement or membership to a particular group), which is then to be predicted with the aid of explanatory or input variables.

Because the research question asks how artificial intelligence can support the phase of screening application documents, especially letters of motivation/cover letters, one method from the supervised as well as one from unsupervised class will be used: cluster analysis combined with multiple linear regression and neural networks will be both applied for performance comparison. In order to verify the performance, the following test hypotheses were established, which seem reasonable from the perspective of a HR recruiter, if he or she should be supported by an AI.

The first hypothesis is that the cluster analysis from the class of unsupervised learning will find applications that are close to the job advertisement (this can be tested if the job advertisement is part of the overall dataset). Here, the assumption is that applicants relate their story telling in the letter of motivation/cover letter to the language of the job posting. The more an applicant can relate his or her own history to the language used in the job posting, the closer it is to that. The essential analysis problem in this case is the radical exception that an applicant would only copy the advertisement text into the letter of motivation/cover letter—thereby, he or she would have 100% coverage, but would disqualify him or herself in the context of a real application. This means that a letter of motivation/ cover letter should not have 100% coverage with the job posting (not equal to a distance of 0).

The second hypothesis is that cluster analysis can replicate the ten judgment steps of the overall scale. Here, the initial consideration is that raters make their judgment based on the applicant's word choice. If this hypothesis is true, the placement of applications into clusters is correlated with the overall judgment. Ideally, all applications in cluster A would receive the overall judgment of 10%, all in cluster B would receive 20%, and so on. Since cluster assignment is done via unsupervised learning, it is by no means mandatory that the cluster membership identified via word choice is correlated with the overall judgement. Furthermore, even if there is a correlation, the clusters are not ordered; for example, belonging to cluster 2 does not mean that the overall verdict is 20% on average. Moreover, cluster analysis (K-means) always randomly determines a starting point at each call, then adds documents and searches for new neighbors from their common midpoint. Therefore, the correlation between overall judgment and cluster membership is tested using linear regression (mean R.lm), where the cluster memberships are coded as dummy variables and form the explanatory variables. The advantage of this procedure is that there is no need to order the clusters (such as 2 = 20%); instead, the model fit is used only to test whether the clusters generally have predictive power. Due to the lack of ordering of the clusters, tests for individual parameters (e.g., t-test for cluster 2) will not be meaningful and will also vary per run due to the randomness of the starting point; in contrast, the tests for the overall model and the model fit for the overall model will be stable. For these two hypotheses, multiple linear regression from the class of supervised learning is used in addition to cluster analysis from the class of unsupervised learning.

The third hypothesis is that cluster analysis can simplify the recruiter's work and suggest the optimal level of clustering. Here the assumption is that the proposed 10% judgment steps are not really used [40], but classically, as with Likert scales, human judgment tends to form 3, 4 or 5 differentiations. For the HR recruiters, a positive result on this hypothesis would also be meaningful if the first two hypotheses cannot be proven true. If the clusters determined have a high predictive power, the HR recruiter would only have the necessity to check the quality of one application per cluster. Thus, he could immediately draw conclusions about the quality of all applications within a determined cluster. For this hypothesis, in addition to the cluster analysis, the results of the multiple linear regression are also used, so that methods of both classes are applied.

The fourth hypothesis includes an alternative method of supervised learning, which is neural network analysis. It postulates, that the application of a neural network leads to good model results. Accordingly, the fifth hypothesis is that the neural network for the combined data set should even lead to better results than the linear regression of the optimal cluster model. This hypothesis follows from the consideration that the neural network was trained as a variant of supervised learning for the prediction of the response variable, whereas the structures revealed by clustering need not necessarily be correlated with the response variable. However, with the neural network there is also an increased risk of overfitting, especially with small datasets. To eliminate this possibility, the training data can be randomly assembled in several runs.

Unfortunately, since sets 1–3 are very small, hypotheses one to four are carried out for all sets including the combined set—however, hypothesis five can only be tested with the combined set.

10.5   The Results and Discussion

The analysis is presented below based on the previously established hypotheses for each of the sets and the combined set—in the graphs, set 1 is always at the top left, then set 2 at the top right, followed in the next row by set 3 and then the combined set.

Starting with hypothesis 1 that the cluster analysis is able to identify applications close to the job posting (0STL), the following dendrograms can be used to illustrate the distances (cf. Fig. 10.2).

For set 1, the closest application to the job posting is one application with the abbreviation UZUI, whereas for set 2, it is a whole cluster (consisting of YIKO, GGOH, AIOW, ... OPST, RCNB). For set 1, the job posting as well as the application forms a cluster, while in set 2 the job posting alone forms a cluster with a second cluster nearby. In this respect, the diagrams of the cluster analyses already provide indications that in set 1 there is at least one application that has a high similarity to the job advertisement, while this is not the case in set 2. One possible reason for this observation here is the quality of the applications

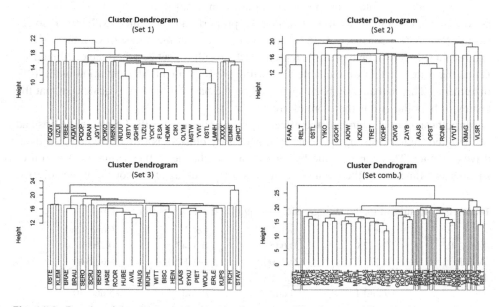

Fig. 10.2 Results of the cluster analysis—dendrograms of the data sets (distances)

from set 2, as the applications probably tend to correspond to standard cover letters in that the students have made little effort to adapt their cover letter to the advertisement. Another reason could be that the application which is part of the cluster with the job posting is too close to the text of the job posting and should therefore be disqualified.

In set 3 it is clearly visible that two applications (FICH/STAV) lie apart from the rest of the data; these two applications would form their own cluster even with a fairly high cut (approx. Height 23 in the diagram), while the remaining documents formed the second remaining cluster. In set 3, the job advertisement forms its own separate cluster, too. In this respect, a weaker connection between the letters of motivation/cover letters and the text of the job posting is also evident compared to set 1.

Whereas in the case of set 3, it is reasonable to assume that the commitment is lower in the case of a minor subject and therefore a standard cover letter was uploaded. In the case of set 2 where the context of a major in HRM is given, it is only reasonable to assume that the students either did not know how or did not want to perform better.

The discussion of hypothesis 2 also refers to Fig. 10.2. Here we can look at the red frames of the applications that correspond to the ten identified clusters. What is now relevant is the extent to which this cluster membership is correlated with the overall judgment. For this purpose, the model goodness of fit and the associated test for overall explanatory power of the regression model (cf. Table 10.2) are used.

Here it can be seen that in set 1 and 2 a relatively high proportion of the variance in the overall judgments (Multi. $R^2 > 60\%$) can be explained by cluster membership. The F-test for the overall explanatory power of the respective regression model shows that the explanatory variables of the model (= cluster membership) are jointly statistically significant for set 1, with an acceptable error ($p < 6\%$), but not significant ($p > 25\%$) for set 2. For set 3 as well as the combined set, it is clear that the proportion of variance explained is lower (Multi. $R^2 < 35\%$); consistent with this observation, cluster membership variables are not significant ($p > 50\%$). One possible reason for the result for set 2, 3, and the combined set is that the student raters were unable or unwilling to implement differentiation in the 10% increments. This had already been suspected on the basis of the Likert scales, so that the testing of hypothesis 3 is therefore of great interest.

Hypothesis 3 now examines the effect of the number of specified clusters, i.e. instead of the fixed number of 10, the number of necessary clusters is to be revealed in a data-driven manner. This can be done by looking at the change in the within-cluster sum of squares as a function of the number of clusters or by looking at the course of the adjusted

Table 10.2 Results of the linear regression—R^2/p/df of the data sets (for k = 10)

	Set 1	Set 2	Set 3	Comb. Set
Multi. R^2	0.6173	0.6453	0.3347	0.2536
Adj. R^2	0.3713	0.2400	−0.0454	−0.0540
P	0.05961	0.2768	0.5554	0.6428
Df	9 and 14	8 and 7	8 and 14	8 and 30

Note: All values cut off after 4 digits (no rounding)

R^2 as a function of the number of clusters. In contrast to the ordinary R^2, the calculation of the adjusted R^2 takes into account the different numbers of clusters (= number of explanatory variables), so that it can be used well for model selection with different numbers of explanatory variables.

An evaluation of the trend of the adjusted R^2 (cf. Fig. 10.3) leads to the following conclusion: While the results for set 1 continue to clearly indicate the performance of method piloting, reducing the set of optimal clusters to 5, the results for the other sets show either decreasing trends of adjusted R^2 as more clusters are added (set 2), or even low adjusted R^2 around zero from the beginning (set 3, combined set). This significant difference in performance may have arisen in the following places: (1) Applicants in set 1 had a job available that matched their program profile and major, while applicants from set 2 and set 3 had a generic job posting. (2) The applications from set 3 were in the context of a minor and therefore likely to be more general in nature and less focused on the actual job advertisement in terms of content. Therefore, the evaluation of set 2 is exactly in between that of set 1 and set 3. (3) Of course, as previously mentioned, a rather low precision was to be expected simply due to the small sample size in the context of this pilot study. Therefore, these results should not yet be taken as a refutation of the methodology. (4) However, it cannot be ruled out in principle that cluster membership is not systematically correlated with overall judgement. This question must ultimately be answered with a larger data set. Nevertheless, based on the analysis, the results with the best adjusted R^2 can be extracted:

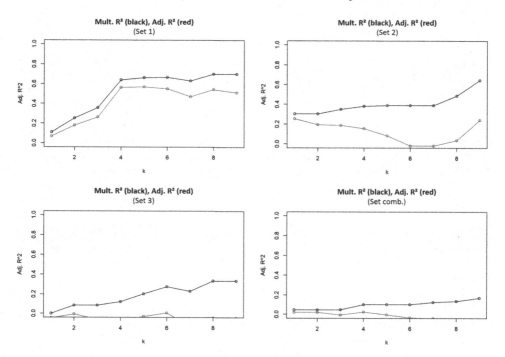

Fig. 10.3 Results of the cluster analysis—course of the adj. R^2 in dep. of the cluster number

This is the case when set 1 is examined with 4 clusters, set 2 with 3 clusters, set 3 with 6 clusters, and the combined set with 4 clusters (cf. Fig. 10.4).

Although the quality of the division into different clusters for these data sets can only be assumed to be average in one case (set 1 ~ 35% of the variance, set 2 ~ 20%, set 3 ~ 0%, comb. set ~0%), this is still a promising result. If further studies with real applications confirm the results of set 1, then a HR recruiter could indeed let the cluster analysis generate a sorting and only one application per batch have to be checked for identifying the quality of the cluster.

The fourth hypothesis assumed that the application of a neural network should lead to good results. The application of a neural network is defined by the set of input variables (these are given by the set of words used in the DTM matrix), the set of output variables (here only 1 for the overall judgment) and the set of hidden neurons in the network. For this analysis, a simple single-layer neural network was chosen. The set of hidden neurons must be between 1 and X, where the X is to be chosen by the set of input data cases divided by 5 [41]. Since the inner activation functions of the neural network are designed to have values between 0 and 1, the input and output data must be normalized. If one then iterates over the number of hidden neurons, the error (RMSE) between the determined output values and the real values can be output (cf. Table 10.3).

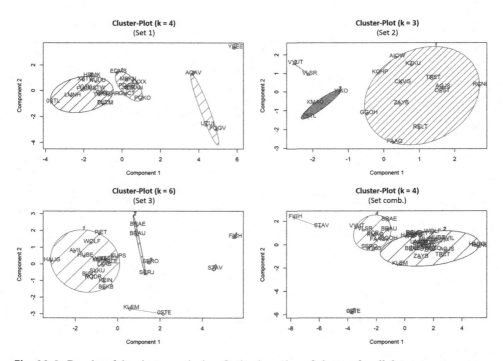

Fig. 10.4 Results of the cluster analysis—Optimal number of clusters for all data sets

Table 10.3 Results of the neural network—RMSE depending on the number of hidden neurons

	Set 1	Set 2	Set 3	Comb. Set
H max.	5	3	5	8
H = 1	0.1382	0.2084	0.2531	0.1478
H = 2	0.1331	0.1117[a]	0.1301	0.0756
H = 3	0.1188	0.2526	0.1544	0.0869
H = 4	0.0991[a]	–	0.1310	0.1124
H = 5	0.1320	–	0.0682[a]	0.0813
H = 6	–	–	–	0.0584
H = 7	–	–	–	0.0777
H = 8	–	–	–	0.0578[a]

Note: All values truncated after 4 digits (no rounding)
[a]Min

Table 10.4 Results of cluster analysis/linear regression against neural network (RMSE)

	Comb. Set
Lin. Reg. with optimal cluster number (k = 4)[a]	0.2380
Neural network with optimal number of hidden neurons (H = 4)[b]	0.1507 (training)
	0.2552 (test)

Note: All values truncated after 4 digits (no rounding)
[a]Single run
[b]Mean values for 10 runs

Here it can be seen that the minima (a) of the different sets vary between 0.0578 and 0.1117. This corresponds to an error of 6–11% for the normalized values. Accordingly, a valid result can be obtained with this method not only for set 1, but also for sets 2, 3 and the combined set. The problem, however, is that this involves what is known as overfitting—the network learns the data set "by heart", but cannot be used to make judgments about other data. While this hypothesis can be considered confirmed, the performance comparison of the methods is still pending.

To answer hypothesis 5, that the neural network for the combined data set should lead to better results than the linear regression of the optimal cluster model, bootstrapping must be used for the neural network—this could also be done for the cluster analysis. Here the combined set was split ten times into training and test data and then the RMSE was stored. The mean RMSE can then be output over all ten iterations. Although only the R^2 values were previously extracted for the regression, the RMSE can of course also be determined, so that both results/methods can be compared with each other (cf. Table 10.4).

It can be seen from presented table that the results (RMSE) of the cluster analysis (0.23) are indeed outrun by the neural network for the training data set (0.15). At the same time, the results for the test subset are slightly worse (0.25) than those of the cluster analysis. Again, it can only be pointed out that the data set is too small for a final evaluation of the method. At the same time, the ten percentage points better result of the model error in the case of the neural network gives hope that supervised learning will ultimately lead to valid results for this task as well.

10.6   Outlook

The goal was to analyze how artificial intelligence can support the screening phase of the staffing process, in particular for the analysis of application documents when using (un) supervised learning.

A final assessment of the results of the hypotheses shows that even in small and medium-sized companies with a small amount of data, it is possible to obtain results using methods of unsupervised learning (cluster analysis). Set 1 promises a relatively moderate goodness of fit, if the problems of sets 2 and 3 can indeed be explained by the lower motivation to perform. Accordingly, cluster analysis could be used to identify the optimal set of clusters so that subsequently, as a HR recruiter, only 1 or 2 applications per cluster need to be looked for quality control. Since a neural network could not achieve far better results (under the condition of small dataset) than cluster analysis, this particular method seems to be rather an "overkill" for small datasets for this question. On the other hand, for larger companies with larger datasets it would be advisable to perform this performance comparison again and validate the results.

However, the results of this method piloting are based on imperfect data. Therefore, it is advisable in any case to compare these results once again with a data set generated under real application conditions, where the applicants have a high level of self-motivation to achieve an invitation which is reflected in the quality of the documents. At the same time it must be remembered that for this analysis, only the overall judgment of the cover letters from the recruiters' perspective were used. It is, of course, questionable whether the cluster analysis leads to better results if the formal aspects or the linguistic criteria are assessed separately or even additionally. This also applies to the use of the neural network.

By restricting the research question to the screening phase of the staffing process and in particular to the analysis of letters of motivation/cover letters, one can probably count on better prediction results for more complex analyses and better data. Here, on the one hand, the resumés and references/certificates were not taken into account in the text-based analysis; at the same time, optical evaluations with convolutional networks are also available. Especially for these extensions of the research question, however, larger data sets with more than 100 applications for one job posting are necessary.

References

1. Deutscher Bundestag (2018) Enquete-Kommission zur künstlichen Intelligenz eingesetzt. Von https://www.bundestag.de/dokumente/textarchiv/2018/kw26-de-enquete-kommission-kuenstliche-intelligenz/560330. Accessed: 17. Febr. 2022

2. Kaplan A, Haenlein M (2019) Siri, Siri, in my hand: Who's the fairest in the land? On the interpretations, illustrations, and implications of artificial intelligence. Bus Horiz 62(1):15–25
3. Jeffrey JA et al (2018) The future of jobs report 2018. World Economic Forum, Genever
4. Jeffrey JA et al (2016) The future of jobs—employment, skills and workforce strategy for the
fourth industrial revolution. World Economic Forum, Genever
5. Daheim C, Wintermann O (2015) 2050: Die Zukunft der Arbeit. Ergebnisse einer internationalen
Delphi-Studie des Millennium Project. Bertelsmann Stiftung, Gütersloh
6. Kröse B, van der Smag P (1996) An introduction to neural networks. University of Amsterdam,
Amsterdam
7. McCulloch W, Pitts W (1943) A logical calculus of the ideas immanent in nervous activity. Bull
Math Biophys 5:115–133
8. Bishop CM (1996) Neural networks for pattern recognition. Clarendon, Oxford
9. Kaastra I, Boyd M (1996) Designing a neural network for forecasting financial and economic
time series. Neurocomputing 10(3):215–236
10. Kohonen T (1995) Self-organizing maps. Springer, Berlin
11. Zhang PG, Qi M (2005) Neural network forecasting for seasonal and trend time series. Eur J
Oper Res 160(2):501–514
12. Kosinski M, Bachrach Y, Kohli P, Stillwell D, Graeoel T (2013) Manifestations of user
personality in website choice and behaviour on online social networks. Mach Learn
95(3):357–380
13. Kosinski M, Stillwell D, Graepel T (2013) Private traits and attributes are predictable from digital records of human behavior. Proc Natl Acad Sci USA 110(15):5802–5805
14. Hu M, Zhang G, Jiang C, Patuwo B (1999) A cross-validation analysis of neural network out-of-
sample performance in exchange rate forecasting. Decis Sci 30(1):197–216
15. Vellido A, Lisboa PJ, Vaughan J (1999) Neural networks in business: a survey of applications
(1992–1998). Expert Syst Appl 17(1):51–70
16. Widrow B, Rumelhart DE, Lehr MA (1994) Neural networks: applications in industry, business
and science. Commun ACM 37(3):93–105
17. Wong BK, Bodnovich TA, Selvi Y (1997) Neural network applications in business: a review and
analysis of the literatue (1988–1995). Decis Support Syst 19(4):301–320
18. Wong BK, Lai VS, Lam J (2000) A bibliography of neural network business applications
research: 1994–1998. Comput Oper Res 27(11–12):1045–1076
19. Dengler K, Matthes B (2018) Wenige Berufsbilder halten mit der Digitalisierung Schritt. IAB-
Kurzbericht 4:1–11
20. Dengler K, Matthes B (2015) Folgen der Digitalisierung für die Arbeitswelt:
Substituierbarkeitspotenziale von Berufen in Deutschland. IAB-Forschungsbericht 11/2015
21. Borgert S, Helfritz KH (2019) Künstliche Intelligenz in HR. DFGP e. V., TU Kaiserslautern,
Kaiserslautern
22. Fichtner U, Fischer S, Michael A, Neyer A-K (2019) Künstliche Intelligenz: Probieren, lernen,
machen. In Schwuchow K, Gutmann J (Hrsg) HR-Trends 2020: Agilität, Arbeit 4.0, Analytics,
Prozesse. Haufe, Freiburg, p 282–290
23. Strohmeier S, Piazza F (2015) Artificial intelligence techniques in human resource management—a conceptual exploration. In Kahraman C, Onar SÇ (Hrsg) Intelligent techniques in engineering management. Springer, Cham, p 149–172
24. Roedenbeck MR (2020) Die richtigen Fragen stellen. Z OrganisationsEntwicklung 39(1/20):64–69
25. Garvin DA (2013) How Google sold its engineers on management. Harvard Bus Rev 91(12):74–82
26. Bakir D (2019) Wenn Ihr Chef weiß, dass Sie kündigen werden, bevor Sie es selbst wissen. Von
https://www.stern.de/wirtschaft/job/kuendigung–wenn-die-ki-vor-ihnen-weiss–dass-sie-kuendigen-
werden-8668660.html. Accessed: 17. Febr. 2022

27. Bauer T, Erdogan B, Caughlin D, Truxillo D (2020) Human resource management—people, data, and analytics. Sage, Thousand Oaks
28. Gopalakrishna ST, Varadharajan V (2019) Automated tool for resume classification using semantic analysis. Int J Artif Intell Appl 10(1):11–23
29. Deng Y, Lei H, Li X, Lin Y (2018) An improved deep neural network model for job matching. In: 2018 international conference on artificial intelligence and big data. IEEE, p 106–112
30. Geetha R, Bhanu SR (2018) Recruitman through artificial intelligence: a conceptual study. Int J Mech Eng Technol 9(7):63–70
31. Fischer S, Michael A, Fichtner U (2019) Zwischen Euphorie und Skepsis—KI in der Personalarbeit. Bundesverband der Personalmanager, Berlin
32. R Core Team (2019) R: A language and environment for statistical computing. R Foundation for Statistical Computing, Vienna, Austria. Quelle: https://www.R-project.org/. Accessed: 17. Febr. 2022
33. Wickham H, François R, Henry L, Müller K (2020) dplyr: a grammar of data manipulation. R package version 0.8.4. Quelle: https://CRAN.R-project.org/package=dplyr. Accessed: 17. Febr. 2022
34. Topping KJ (1998) Peer assessment between students in colleges and universities. Rev Educ Res 68:249–276
35. van Zundert M, Sluijsmans D, van Merrienboer J (2010) Effective peer assessment process: research findings and future directions. Learn Instr 20:270–279
36. Feinerer I, Hornik K (2019) tm: text mining package. R package version 0.7–7. Quelle: https://CRAN.R-project.org/package=tm. Accessed: 17. Febr. 2022
37. Feinerer I, Hornik K, Meyer D (2008) Text mining infrastructure in R. J Stat Softw 25(5):1–54. Quelle: http://www.jstatsoft.org/v25/i05/. Accessed: 17. Febr. 2022
38. Frochte J (2019) Maschinelles Lernen—Grundlagen und Algorithmen in Python. Hanser, München
39. Lantz B (2019) Machine learning with R: expert techniques for predictive modelling. Birmingham, Packt
40. Boone HN, Boone DA (2012) Analyzing Likert Data. J Ext 50(2):2TOT2
41. Klimasaukas CC (1991) Applying neural networks, Part 3: trainning a neural network. In: Proceedings in artificial intelligence, p 20–24

Prof. Marc Roedenbeck received his doctorate on individual path dependency and organizational consulting from the Free University of Berlin. Before moving to Technical University of Applied Sciences Wildau, he worked as a senior (HR) consultant in a global management consulting firm as well as a regional bank in Berlin, as commercial director of a Leibniz research institute, as branch manager of a management headhunting agency in Berlin and Hamburg, and as founding dean for the faculty of business in a private university.

Salmai Qari received his PhD from the Berlin Doctoral Program in Economics & Management Science (BDPEMS) at Freie Universität Berlin. Among other things, he worked in the field of statistical modeling in the banking sector. Before moving to the professorship of econometrics at the HWR Berlin, he was a senior research fellow at the Max Planck Institute for Tax Law and Public Finance and also offered courses at the LMU Munich.

Mr. Marcel Herold has been a research assistant at the Wildau University of Applied Sciences since September 2019. He is doing his doctorate on the topic of "Application of artificial intelligence in recruiting"

An AI-Based Framework for Speech and Voice Analytics to Automatically Assess the Quality of Service Conversations

Mathias Walther

Abstract

In this chapter, an innovative two-stage classification framework is presented that can predict quality-inducing criteria in call center conversations with explainable rules based on multiple models for speech expression. Through this basic classification, a symbolic representation of the speech expression is generated that is both understandable to experts and can be processed by classification algorithms. In the second stage, learning procedures are used to combine the recognized speech and voice features into a classification of quality factors. Rules and decision trees map the functional relationships to the relevant features and can thus explain the perceived quality factors on the basis of the recognized speech-voice features.

11.1 Introduction

11.1.1 Motivation

In recent years, the importance of call centers is increasing more and more. In 2013, there were already 6800 call centers with a total of 520,000 employees making 25 million calls per day [17]. Since then, the service market in Germany has grown steadily at about 7% per year. In the market, there is a high pressure to succeed from clients as well as a high competitive pressure. Thus, high call quality is one of the most critical success factors in

M. Walther (✉)
Technical University Wildau, Wildau, Germany
e-mail: mathias.walther@th-wildau.de

T. Barton, C. Müller (eds.), *Artificial intelligence in application*,
https://doi.org/10.1007/978-3-658-43843-2_11

modern call centers, since the communicative work of agents at the main customer inter-face is the core process of a call center. All conversations that are conducted on a daily basis should be evaluated in the course of quality management and agents should be made aware of opportunities for improvement. Mostly, shadow monitoring is used for this pur-pose, which refers to a direct listening of the agent's conversations at the workplace [7]. However, this is very time consuming and therefore only possible for a small percentage of conversations. In order to support the work of conversation coaching, which is an essen-tial part of quality management in call centers, more advanced analyses of the conversa-tion based on prosodic features of speech, i.e., features related to emphasis, volume, and pauses, as well as analyses of semantic features of the dialogue are required [6]. Currently used software solutions only provide statistical analyses of general conversation character-istics such as the length or the number of accepted conversations. In order to support the coaching process, which plays a significant role in the quality of service in the call center, it should be possible to automatically measure and analyze the call quality. The software should be able to record (speech) patterns in call center conversations based on speech science methods, evaluate the conversations in real time, and visualize the speech data obtained in this way. In the future, this technical innovation should make it possible to classify conversations conducted in call centers according to scientifically proven quality criteria, thus making a major contribution to increasing the quality of service in call cen-ters. The possibility of an automatic evaluation of factors that have an influence on the perceived call quality is thus an important step towards ensuring consistently high quality and increasing the competitiveness of the call center. There are two main reasons for auto-mating the quality assurance process:

11.1.1.1 No Objective Quality Assessment

Due to the lack of scientific discourse in the call center industry, there are no uniform, reli-able definitions of call quality.

11.1.1.2 No Explainable Procedures and Characteristics

Automatic support for agents and trainers or team leaders is only effective if the underly-ing decisions are comprehensible. On the one hand, methods with a good classification performance cannot justify their decisions due to their internal structures; on the other hand, signal features extracted by complex mathematical methods can only be provided with semantics in a few cases. Therefore, the complex task of automatic introspectable call quality assessment can only be solved with the combination of classification models trained on different tasks.

11.1.2 Interdisciplinary Research Approach

The idea of automatic quality assessment arose in the interdisciplinary research project "Research and Optimization of Call Center Communication", which is conducted under the direction of the Seminar for Speech Science and Phonetics at Martin Luther University with various partners [10]. The object of research is the analysis of the basic processes of a conversation, which are important for call centers, with the aim of improving communication between customers and agents. For this purpose, it must first be determined by which characteristics a good conversation is described. Call center conversations have special characteristics in the variation of speech expression. In professional telephone conversations, a discrepancy between routine and individuality can often be observed. On the one hand, the customer has the expectation of being advised individually. On the other hand, the agent's actions are characterized by a high degree of routine. In addition to the speaking-voice aspects, the rhetorical design of the conversation is an approach to increase the quality of the conversation. From the determined criteria, which describe the quality of a conversation in a speaking-voice, rhetorical and argumentative way, scientifically founded training concepts are created. These take into account both the situational characteristics of oral communication in the call centre context and the agent's speech training. Professional communication on the telephone is a great challenge for the parties involved, since only a small part of the communication possibilities can be used via the audio channel. The human communication repertoire includes auditory and visual components, such as speech movements, facial expression or posture of the speaker [18]. On the phone, communication has to do without the visual component, which makes it difficult to assess the partner. Content analyses also require a complete transcription of all conversations under investigation, which requires very powerful text recognition. In the research project "Research and Optimization of Professional Telephony" the analysis of speech and voice characteristics is in the foreground. Therefore, speech recognition and processing are not considered in detail, as well as in most research papers, e.g [24]. The consensus of the research is to focus the work on the recognition and classification of paralinguistic features. These exist separately from linguistic content and include speaker characteristics such as physiology, emotions, speech habits, health, and dialect [24]. Research in this area is highly interdisciplinary. Information systems research, as a formative science, is responsible for designing new procedures and integrating them into application scenarios. From the technical and methodological perspective, approaches to automatic quality control are developed and discussed. The results presented here are based in excerpts on individual contributions by the author [31–33] and his dissertation [29] with the aim of presenting the concepts with a focus on the application of the methods from business informatics.

11.2   State of Research

11.2.1 Use of Voice Technology in the Call Centre

Some application scenarios have been designed for the use of speech technology in the last 25 years. The goal of the current development in call center technology is to automate and accelerate business processes through speech technologies [11]. Basically, all applications of speech technologies in call centers can be divided into two functional areas—assistance and monitoring. The interactivity of the conversation requires the agent to quickly grasp the customer's concern and resolve it to the customer's satisfaction. Assistance systems support the agent in quickly identifying and solving the problem. Monitoring, on the other hand, is mainly for the management of the call center or the team leaders. Information can be collected about the nature of customer queries and their solutions [15].

For some time now, so-called IVR (Interactive Voice Response) systems have been used as assistance systems for callers. IVR systems are dialog systems with which the caller can communicate while on hold. Ususally they are controlled by keyboard input or by speech [7]. Emotion recognition at IVR is a scenario of using technologies of automatic analysis of paralinguistic features and recognition of speaker characteristics that has been considered since the mid-1990s. Here, emotion recognition can serve several purposes. For example, by classifying the caller into emotional categories, angry customers can be routed directly to an agent [34]. Furthermore, recorded conversations can be emotionally classified and used as examples for training purposes [35]. First experiments on the classification of basic emotions in realistic telephone recordings are described by Petrushin 1999 [20]. Inbound conversations are categorized in terms of their emotionality and prioritized for the processing procedure. For the study, a corpus consisting of 56 played phone calls was included. "Emotional" and "calm" were chosen as classes to be recognized. Recognition rates of about 75% were obtained in these experiments [20]. Similar values are also reported in other studies [4, 12].

Support systems in call centers should also have the ability to process text by understanding natural language [11]. To this end, various authors present prototype systems that combine speech recognition and speech processing of the written text for semantic analysis [15]. These applications are also referred to as speech analytics [8]. A subfield of Speech Analytics is phonetic search, where in audio databases, in contrast to speech recognition, the search is not performed in the transcribed text, but directly using phonetic patterns in the speech recording. Advantages of phonetic search are higher speed compared to speech recognition and independence from predefined vocabulary [8]. One application for speech analytics and speech recognition systems is directory assistance. A study compared how well automated systems and humans can perform different tasks. A success rate of 55.2% for automatic systems was found for recognizing the city and state of the United States. Agents were able to perform this task with 100% success rate [3].

11.2.2 Automatic Detection of Paralinguistic Features

Work from the field of speech signal processing serves as a starting point for the methods described here, since it is a first approach to the classification of speech effects by machines. This conventional one-stage classification is based on the signal level. Figure 11.1 shows the signal level for two utterances belonging to exemplar classes "good talk" and "bad talk". The upper part of the figure shows the digitized signal, the lower part the spectrogram. This is a combined time-frequency domain representation and is very suitable for illustrating the individual spectral components of the speech signal. Using the frequency decomposition, spectral attributes can be calculated over all signal sections to be analyzed. These are primarily mathematical figures and poorly interpretable acoustically, such as the Mel-Frequency Cepstral Coefficients (MFCC). Some features are easily interpretable, e.g. the fundamental frequency (F0), which can be simplified as the pitch of the voice and is a central element of speech science analysis [21]. In Fig. 11.1, the extracted fundamental frequency trajectories for both utterances are marked. The "good" conversation is characterized by a uniform melodic progression, whereas the speech melody in the "bad" conversation is very agitated at the beginning. This progression can be captured very easily with statistical measures such as the difference between minimum and maximum. Thus, a simple rule could be defined for the example that separates "good" and "bad" conversations. Since the process of speech perception is very complex and, in contrast to the example shown, the classes are not so easy to separate in the normal case, additional attributes must be used and generalized rules derived from them that apply to as large a number of audio segments as possible.

11.2.3 Challenges

The description of motivation and fundamentals in the previous sections show that paralinguistic feature detection is a multifaceted topic. In recent years, many approaches have

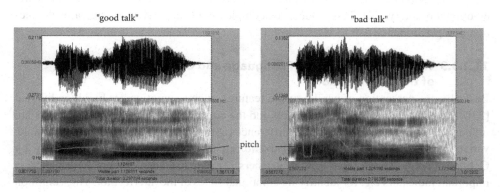

Fig. 11.1 Simplified characteristics

been presented to solve fundamental problems. The detailed analysis of the state of the art of automatic paralinguistic feature detection identifies key challenges that can be summarized in the following three questions [29]:

11.2.3.1 What Measure Defines Good Detection?

Most pattern recognition work is technically driven and aims at mapping the (classification) problem as well as possible and improving the recognition rates more and more. When comparing the results of different experiments, it must be noted that the assessment of the classification quality is crucially dependent on the task. For example, while a hit rate of 75% for a speech recognition task with a limited vocabulary cannot be described as "good", a personality recognition classification task solved with 75% correct assignments is definitely "good" [24].

11.2.3.2 How Can the Properties to Be Recognized Be Defined and How Can This Definition Be Consistently Mapped to a Corpus?

In contrast to the classical applications of pattern recognition, there are no absolute facts in the classification of speaker features, that is, a priori certain classes to be recognized. Of the paralinguistic features, only gender, age and, with limitations, alcoholization can be validly determined and measured. The automatic systems can therefore only reflect the opinions and impressions of a few people who evaluate the data set.

Developing emotion recognition systems to process real live speech data is difficult with traditional emotion databases [12]. These usually contain acted emotions or are not representative of call center conversations for other reasons. One challenge in using real data, i.e., recorded conversations, is the usually uneven distribution of classes. Studies show that a large proportion of calls are neutral, while the small proportion of angry callers (3–5%) need to be detected [16]. Besides the acquisition of real data, the number of states or classes to be distinguished is also crucial for potential applications in call centers. The complexity of the task increases with the number of features to be recognized. When distinguishing five emotions, only 56% recognition rate is achieved [28]. Therefore, in general, for emotion recognition at IVR, it is sufficient to distinguish two classes: neutral and emotional or angry [22]. This reduces the complexity of the model and increases the recognition performance.

11.2.3.3 What Is the Influence of Language and Content of the Conversation?

Basically, the perception of paralinguistic features, such as emotions, is highly individual [23]. The perception of prosodic features and the attribution to speaker characteristics are influenced by cultural and linguistic differences. Studies have shown that, for example, variation in speaking voice pitch related to friendliness is interpreted differently by German, French, Greek, and Turkish listeners [2]. For listeners, content information is important indicators e.g. for identifying emotional states of the interlocutor. Thus, emotion recognition is more difficult for listeners in unfamiliar languages than in familiar ones

[27]. Therefore, when analyzing experiments, it must be taken into account that existing algorithms only process the audio signal and do not use semantic information.

11.3   Design of an Intelligent Assistant to Improve Call Quality

11.3.1 New Approaches to Assessing Call Quality

Outbound telephone sales calls have a bad public image. Sales activity is often perceived as exhausting, manipulative and even damaging to reputation. One reason for the negative image is the routine of conversation and argumentation caused by the high volume of calls. The conversations conducted again and again according to the same pattern lead to routine actions on a conversational and vocal level, which are perceived by the customer. Neuber and Hirschfeld (2013) [10] refer to this as "industrial conversation production". Current research results lead to the conclusion that the communication work done in call centers is largely poor. In addition to repeated service such as long waiting times and pushy behavior of agents, this is also caused by a lack of competence as well as unmotivated and phrase-like conversation [17].

The quality of the conversation is therefore a central factor for customer satisfaction. For its part, customer satisfaction is strongly dependent on the relationship satisfaction that develops between agents and customers in the course of the conversation. The outcome and quality of conversations are largely influenced by conversation-shaping skills as well as interpersonal interaction. However, the effect of the agent's personality cannot be fully appreciated through standardization such as scripted conversation [13]. In summary, the lack of scientific foundation of both communication processes from a speech science perspective and training concepts can be identified as the main obstacle in the objective evaluation of conversations in professional telephony. For the measurement of call quality, only few works and no universally valid assured definitions exist so far. Therefore, a catalogue of criteria was developed for speech science investigations in the research project, which for the first time enables the description of quality features by well-founded evaluation criteria [13]. For speech science investigations in the research project, six phonetic and rhetorical quality factors were identified by a group of experts after extensive literature research and qualitative analysis of 840 authentic call center conversations. These factors are used to characterize outbound sales calls and are presented in Fig. 11.2. Quality factors and the phonetic-voice design were divided into evaluation criteria that allow for description. Six criteria of quality were selected for the study (Conversational partner orientation, Personality/Authenticity, Situational adequateness, Conversational form, Emotionality, and Intelligibility). The focus of the analysis is on the speech expression, since this can be measured qualitatively and partly by means of acoustic correlates and is responsible for the speech effect. Speech expression includes, among others, the evaluation criteria perceived pitch, loudness, timbre, speaking rate as well as the feature complexes accentuation, structuring, rhythm and speech tension. The complete evaluation

Fig. 11.2 Factors of
call quality

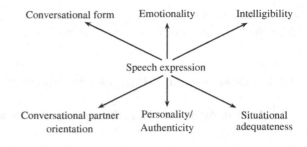

Fig. 11.3 System vision of an
interactive system for call
evaluation

catalogue can be found in [13]. The description of the speech expression is based on the feature catalogue of Bose [1]. The detailed description will be omitted here.

11.3.2 System Vision

The system vision in Fig. 11.3 outlines an interactive decision support system that evaluates conversation quality in real time and shows opportunities for improvement based on speech expression. By providing detailed feedback, the agent can adjust his or her speaking style and improve the overall rating or perceived call quality. The conversation monitor on the right side of Fig. 11.3 exemplifies two criteria of speech expression as well as the overall rating of the conversation and the perceived competence as a factor of conversation quality. Based on the evaluation of the speech expression features, the agent immediately receives concrete suggestions for improvement. In Fig. 11.3, these are, for example, speaking more slowly and taking longer pauses. In order to be able to implement the system's suggestions, the agent must be able to work specifically and precisely with his voice and must know how the system works. The use of the system therefore requires a fundamental redesign of training methods that specifically teach necessary speaking and vocal skills. In addition to conversation support by a real-time system, the following application scenarios for two-stage classification are conceivable [10]:

- In the first step of implementation, an analysis of relevant speech-voice characteristics and quality assessment downstream of the interview can support the training process.
- Based on the real-time evaluation for the agent, a coach or the team leader can be called in specifically in critical conversation situations.
- The knowledge of conversation quality and its modelling can improve speech synthesis systems so that they can read out recurring passages in appropriate and adapted speech to relieve the agent.

11.3.3 Basic Idea of the Framework

The deficits in the theoretical foundation of conversation quality and in the implementation of suitable training methods reveal great potential for automation. However, up to now, characteristics of the conversational partner, such as personality traits or emotions, can only be detected with black box models. These were outlined in Sect. 11.2.2 and have no or only limited ability to justify their classification decisions. As a supporting system in a call centre, classification using black box models without justification is not suitable. Such a system could only mark certain call passages as positive or negative, but not give any hints for improvement. Consequently, the application scenario of non-explanatory classification systems is limited to a filtering function. The research approach described aims to develop two-stage models for the classification of personality attributes and conversational quality. The basic assumption is based on the research result developed within the project that hierarchical cause-effect relationships exist between speech expression and quality attributes. Detailed analyses by expert listeners confirm a relationship between the feature complex of speech expression and perceived quality factors [14]. The idea of two-stage classification is illustrated in Fig. 11.4. The stages are shown here as a "house". The signal, that is, the speech recording in digital form, is found in the "repetition" since the first stage of the classification process is built on the signal level. The speech expression is close to the signal in the lower stage. Based on these correspondences, some speech expression features can be determined relatively reliably by their acoustic correlates [19]. From speech expression to personality, the distance of the factor from the signal increases and at the same time the predictive ability of the models decreases. Personality features are the furthest from the signal. Classification models are therefore worse at detecting them than, for example, emotions or speech expression. These models are also very complex and cannot explain their decisions. The newly developed two-stage classification procedure is represented by the models shown in gray in Fig. 11.4. It provides effect relationships between speech expression and higher quality features such as personality. This allows the targeted investigation of the causes of poor quality conversations.

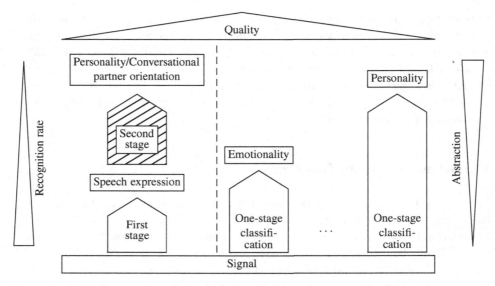

Fig. 11.4 Conceptual model of the two-stage framework

11.4 Experimental Results

11.4.1 Single-Level Classification

All experiments were conducted with an extensive corpus that was created within the framework of the project according to speech science requirements. The corpus is based on real sales conversations from three outbound campaigns provided by different call centers. For the experiments, the channels were separated and only the agent's channel was considered. Since full annotation of the conversations was not possible due to limited resources, 218 conversations were selected for the final corpus out of several hundred in the source dataset, which were considered representative with respect to the speech science criteria [13]. The conversations were subsequently annotated by four students of speech science with respect to salience according to the catalogue presented in [13] (see Sect. 11.3.1) using a six-point scale. For the experiments described, these ordinal scaled data were transformed to two and three classes, respectively, in order to apply classification algorithms. Furthermore, random downsampling was used to artificially create an equal distribution of classes per criterion to counteract unequal distribution. From the 48 evaluation criteria recorded in the first version of the catalogue, 19 criteria are used for the experiments described here. In addition to the 13 criteria of speech expression listed in Table 11.1, six important criteria of conversational quality are also used according to Table 11.2. For each criterion, the classes and the total number of instances per class are given in the second column. Except for the accentuation form with three classes, the criteria vary in two classes.

Table 11.1 Recognition performance for criteria of speech expression (one-stage classification)

Criterion	Classes (number of instances)	RR (%)	Algorithm
Loudness	Loud (57), Quiet (57)	96.89	J48
Perceived pitch	High (146), Low (146)	94.84	MLP
Pause type	End pauses (9), Inner pauses (9)	95.00	LMT
Melody jump	Strong (6), Weak (6)	93.00	NB
Pause frequency	High (37), Low (37)	86.43	BN
End melody	Terminal (57), Interrogative (57)	78.79	SMO
Perceived pitch contour	Inflected (120), Uninflected (120)	77.08	MLP
Pause duration	Long (45), Short (44)	75.42	MLP
Accent frequency	High (8), Low (8)	75.00	MLP
Speech rate	Fast (44), Slow (44)	75.00	MLP
Speech tension	High (98), Low (98)	69.97	MLP
Timbre	Pleasant (74), Unpleasant (74)	65.43	BN
Accent form	Dynamic (71), Temporal (71), Melodic (71)	48.00	BN

Table 11.2 Recognition performance for call quality criteria (one-stage classification)

Criterion	Classes (number of instances)	Algorithm	RR (%)
Competence	Incompetent (71), Competent (71)	SMO	78.76
Certainty	Uncertain (55), Certain (55)	MLP	72.73
Kindness	Friendly (73), Unfriendly (73)	NB	67.95
Naturalness	Unnatural (140), Natural (140)	JRip	64.29
Credibility	Unbelievable (85), Believable (85)	BN	58.24
Cooperative	Uncooperative (82), Cooperative (82)	SMO	54.78

For the computation of the signal features, a feature set with 2106 features computed with openEAR [5] is used. The configuration is based on the feature set used for the "Paralinguistic Challenge" of the Interspeech conference 2010 and was extended by the first five formants as well as statistical functionals [26]. In addition to the signal features, the gender of the agent was manually recorded. Classification is performed using eight commonly used algorithms in their implementations from Weka [9]: naive-Bayes (NB), Bayes net (BN), logistic-model-tree (LMT), ripper (JRip), support vector machine (SMO), ada-boost (Ada), C4.5 decision tree (J48) and multilayer perceptron (MLP). As a measure of recognition performance, the mean recognition rate (RR) was obtained using ten-fold cross-validation. Tables 11.1 and 11.2 show an overview of the best classification algorithms with respect to the selected criteria, sorted in descending order of recognition rate. As can be seen in Table 11.1, all dichotomous criteria of the speech expression can be reliably classified with an average recognition rate of more than 65%. The best recognition rates are achieved by loudness and perceived pitch. Here, the good recognition was to be expected since these features are measurable by their acoustic correlates of fundamental frequency, duration and intensity, respectively [21]. The criteria pause type, melody jump and accentuation frequency can also be detected well. However, the data size is small, so the obtained results cannot be considered valid. For accent form, a recognition rate of 48% is achieved, which is significantly higher than the expected value of 33% for three classes.

The call quality recognition performances listed in Table 11.2 range from 55% to 78%, which is better than the statistical expected value of 50%. The Support Vector Machine (SMO) achieves the highest recognition rate of about 78% for competence. However, on average, the quality factors can be recognized worse than the speech expression. The results further show that all classification algorithms provide the maximum recognition rate for at least one criterion. The multilayer perceptron can achieve the maximum mean recognition rates for the studied data in six cases, followed by the Bayesian network with four first rankings. Thus, no global best classification algorithm can be identified ad hoc without statistical analysis.

Due to different approaches to corpora, feature sets and others, direct comparison with work on paralinguistic speech processing is difficult and can only be done roughly. For example, in the 2012 Interspeech Speaker Trait Challenge [25], the benchmark for speaker personality traits (OCEAN or Big-Five) was about 70% recognition rate. These values can also be achieved for competence, confidence and friendliness in the experiments presented here (see Table 11.2). Besides the good classification rate achieved for some criteria, being able to explain classification decisions is of crucial importance for the application scenario described. Of all the conversational quality features, competence is best detected by a support vector machine. However, this classification algorithm does not have good explanatory ability compared to decision tree methods. In general, the explanatory power of signal level models is relatively low due to the large feature vectors.

11.4.2 Two-Stage Classification

Based on the first stage, the second stage models are trained, which exploit the good recognizability of the speech expression. The procedure corresponds to an expert who listens to a conversation segment several times, assigns ratings for the speech expression and draws conclusions for quality features from these ratings. In the first step, each individual instance of the dataset, e.g. competence, is first classified with the basic models for the 13 criteria of speech expression.

Figure 11.5 shows the structure of the procedure. This creates a new instance with 13 calculated features and the original class attribute, i.e. "competent" or "incompetent", in the example "competent". In addition, the gender of the agent is recorded as the 14th feature. Figure 11.6 shows such an instance, following the arff format of Weka.

This transformation from signal representation to symbolic representation aims at making decision tree methods usable for the second stage models. Therefore, the algorithm J48 from Weka is used for these experiments.

With the data sets transformed in this way, the classification trees of the second stage are trained and the classification quality is determined. The results of the recognition rate determined with a cross-validation for the six categories of conversation quality are given

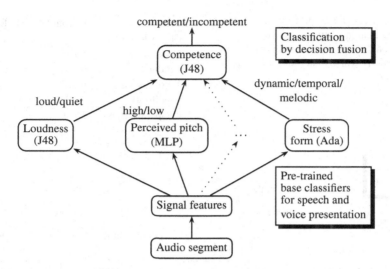

Fig. 11.5 Structure of the second stage model using the example of the quality criterion competence

Fig. 11.6 Training instance
for the second stage competence classifier

```
[...]
@data
loud, low, end pasuses, strong, high,
interrogative, inflected, short, high, fast,
high, unpleasant, dynamic, competent
[...]
```

in Table 11.3. Competence is on average best recognized by the decision trees. This result corresponds with the results of the one-stage models. It can be seen that the recognition of certainty, cooperativeness and credibility is below 60% and thus quite close to the expected value of 50%. Furthermore, the issue of whether the recognition performance of the second stage is sufficient was investigated. For practical use in a system that supports trainers or agents in their daily work by providing automated explanatory ratings of call quality features, the underlying second-stage model must not classify significantly worse than the corresponding single-stage signal model. To validate this, a statistical test was conducted to determine whether the two-stage classification models classified worse than the single-stage equivalents from the previous section. For the criteria safety and friendliness (marked with * in Table 11.3) it can be concluded that they are significantly worse than the one-stage models. Conversely, for the other results, this shows that the models have no disadvantages in recognition quality, but fully exploit the advantage of explanatory power.

Using the rules of the decision trees, which only evaluate speech expression criteria, experts can understand the classification decisions. The learned rules can be flexibly adapted to new data in practical use. As an example, a classification model for competence trained according to this approach is explained below in the form of a decision tree shown in Fig. 11.7. The example was chosen because the tree is particularly compact and informative and also achieves a good recognition rate. With a recognition rate of 77%, the

Table 11.3 Recognition performance for criteria of call quality (two-stage classification), values marked with (*) are significantly worse than those of the first stage

Criterion	Recogntion rate RR (%)
Competence	76.76
Certainty	55.45*
Kindness	61.64*
Naturalness	61.07
Credibility	57.06
Cooperativeness	56.10

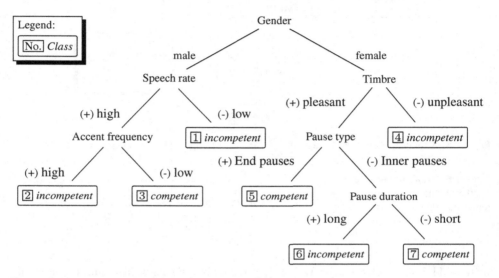

Fig. 11.7 Decision tree for the criterion competence in the second stage

decision tree has almost the same classification quality as the first stage model. The decision trees for the other conversation quality criteria can be interpreted similarly and are described in detail in [29]. In total, the decision tree for competence contains seven rules that can be used to classify unknown conversations. The maximum depth is four for rules 6 and 7, which means that four attributes are considered. The simplest rules 1 and 4 have only two attributes. Rule 1 states that slow talking men are judged as incompetent. Women come across as incompetent with an unpleasant voice (Rule 4). In contrast, they come across as competent with a pleasant voice and noticeable boundary pauses (Rule 5). The examples show that the learned rules are plausible and thus the decision trees can make introspectable decisions.

11.5 Conclusion

The developed approach for the two-stage classification of call quality criteria opens up completely new perspectives of quality management for service providers in professional telecommunications through transparent and explainable evaluations of call quality factors. The introspectability creates great potential benefits for all process participants in the call center. The results show that the two-stage classification methodology is basically feasible and achieves good results in the example of competence. The decision tree shown here has a low complexity and high classification quality. Furthermore, it could be statistically verified that four of the six criteria of conversation quality are not recognized worse with the second-stage decision tree than with the comparable conventional one-stage model. In conclusion, further research needs can be identified. The overall design is based on the assumption that only the best possible recognition of the speech expression enables a good recognition performance in the second stage based on it. In a further test for validation, the important step of choosing the base models was examined. It could be proved that the recognition quality of the base models is crucial for the classification performance of the second stage, i.e. the better the speech expression is recognized, the better the conversation impression can be predicted based on it. Therefore, as a basis for future work, research should be conducted to improve the base classification models.

Due to the influencing factors, the learned rules are only suitable to a limited extent for practical use. Moreover, the two-stage structure of basic model and the aggregated classification in the second stage leads to a stochastic uncertainty of the features due to the classification error of the first stage, which does not occur in conventional classification models. It is further shown that the second stage decision trees are unstable. This means that a small change in the training set leads to a different tree. This is mainly due to the limited data and the different recognition quality of the base models [32]. This instability complicates automation in that it is not possible to work with absolute rules or decision trees that apply to a large number of conversations. It turns out that the generated rules and their classification quality depend on several external factors [30]:

- rhetorical and vocal skills and habits of the agent;
- perceptual habits of the trainer;
- technical circumstances, e.g. when recording the call;
- content of the conversation and the language in which it is conducted.

11.5.1 Outlook: Expert System for Conversation Evaluation

In order to achieve good results with unknown data, the rules must be continuously adjusted so that the influencing factors can be balanced. A systemic framework for the adaptation of rules is an expert system whose concept has been tried and tested for many years. The basic concept of the system is shown in Fig. 11.8, which is based on the general basic components of an expert system [26]. The knowledge base is the main component of an expert system. It contains the domain knowledge and is used for problem solving by logical reasoning. In the proposed system, the knowledge is represented by the learned decision rules and by the knowledge of the experts and trainers formulated in rules. In the problem solving component, the decision rules are

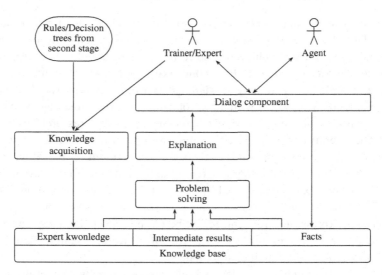

Fig. 11.8 Architecture of an expert system

executed by the rule interpreter. Their functionality is complemented by the explanation component. This describes the inference steps performed by the system to solve the problem. The dialog component is the interface to the user and the expert. Another interface must be created for the trainer. In addition to the feedback function, it must provide a possibility to change rules in order to adapt the system to the environmental conditions. Another interaction with the system takes place in the knowledge acquisition component, since its main function is to transfer knowledge from the knowledge source to the knowledge base. In an application, three usage scenarios for the expert system can be identified. The first scenario describes the initialization of the system. Before it can work, it has to be filled with rules. The initial rule set is generated from the second stage classification rules, as shown in Fig. 11.8. The second deployment scenario is to adapt the rules to the specific situation, especially to the listening habits of the trainer. This is illustrated with the second actor in the role of the expert in Fig. 11.8. In this case, the expert interacts with the knowledge acquisition component. Through the dialogue component, the trainer must be given the opportunity to configure the system to evaluate according to his listening habits. The flexibility of the system is increased by creating multiple user interfaces for different user groups. The third usage scenario is live operation during a conversation, as described in the system vision. The use of increasingly powerful hardware and improved algorithms increasingly allows the processing of speech data in real time. For this to happen, the agent must have the evaluation of the current call reflected back to it in sufficient detail without delay. In live conversation evaluation, the agent works with the dialog component of the system. Based on the real-time evaluation, a trainer or the team leader can be called in specifically in critical conversation situations.

References

1. Bose I (2003) dóch da sin ja' nur mûster: Kindlicher Sprechausdruck im sozialen Rollenspiel. Peter, Frankfurt
2. Burkhardt F, Audibert N, Malatesta L, Türk O, Arslan L, Auberge V (2006) Emotional prosody—does culture make a difference. Speech Prosody 2(5)
3. Chang H (2007) Comparing machine and human performance for caller's directory assistance requests. Int J Speech Technol 10(2):75–87
4. Devillers L, Lamel L, Vasilescu I (2003) Emotion detection in task-oriented spoken dialogues. In: Multimedia and Expo, 2003. ICME'03. Proceedings. 2003 international conference on IEEE, III–549
5. Eyben F, Wöllmer M, Schuller B (2009) openEAR—introducing the Munich open-source emotion and affect recognition toolkit. In: Proc. 4th international HUMAINE association conference on affective computing and intelligent interaction 2009 (ACII2009), Band I, IEEE, p 576–581
6. Focsa I, Neuhaus T (2003) Aufbau eines Qualitätsmanagementsystems im Call Center. GfAH Selbstverlag, Dortmund, p 17–38
7. Fojut S (2008) Call Center Lexikon: Die wichtigsten Fachbegriffe der Branche verständlich erklärt. Gabler, Wiesbaden
8. Gavalda M, Schlueter J (2010) "The truth is out there": using advanced speech analytics to learn why customers call help-line desks and how effectively they are being served by the call center agent. In: Advances in speech recognition. Springer, Berlin, pp 221–243
9. Hall M, Frank E, Holmes G et al (2009) The WEKA data mining software: an update. SIGKDD Explor Newsl 11(1):10–18
10. Hirschfeld U, Neuber B (2011) Optimierungsmöglichkeiten der Telekommunikation aus Sicht der Sprechwissenschaft—Überblick über Fragestellungen und Untersuchungsansätze. In: Hirschfeld U, Neuber B (eds) Erforschung und Optimierung der Callcenterkommunikation. Frank & Timme, Berlin, pp 9–28
11. Lassmann W, Rogge R, Schwarzer J (eds) (2006) Wirtschaftsinformatik: Nachschlagewerk für Studium und Praxis. Gabler, Wiesbaden
12. Lefter I, Wiggers P, Rothkrantz L (2010) EmoReSp: an online emotion recognizer based on speech. In: Proceedings of the 11th international conference on computer systems and technologies and workshop for PhD students in computing on international conference on computer systems and technologies. ACM, New York, (CompSysTech'10), p 287–292
13. Meißner S, Pietschmann J (2011a) Rhetorische und phonetische Einflussfaktoren auf die Qualität von Telefonverkaufsgesprächen. In: Hirschfeld U, Neuber B (eds) Erforschung und Optimierung der Callcenterkommunikation. Frank & Timme, Berlin, pp 215–248
14. Meißner S, Pietschmann J (2011b) Zur Beurteilung der Gesprächsqualität im telefonischen Verkauf—Zwischenbericht über ein Forschungsprojekt. In: Bose I, Neuber B (eds) Interpersonale Kommunikation: Analyse und Optimierung. Lang, Frankfurt, pp 303–312
15. Mishne G, Carmel D, Hoory R, Roytman A, Soffer A (2005) Automatic analysis of call-center conversations. In: Proceedings of the 14th ACM international conference on Information and knowledge management. ACM, p 453–459
16. Morrison D, Wang R, De Silva L (2007) Ensemble methods for spoken emotion recognition in call-centres. Speech Comm 49(2):98–112
17. Neuber B, Hirschfeld U (2013) Sprechwirkungsforschung in der professionellen Telefonie. In: Veličkova L, Petročenko E (Ed) Klangsprache im Fremdsprachenunterricht, Bd. VII. Voronezh State University, Voronezh, p 66–85
18. Neppert J, Pétursson M (1986) Elemente einer Akustischen Phonetik, 2. Aufl. Helmut Buske, Hamburg
19. Paeschke A (2003) Prosodische Analyse emotionaler Sprechweise. Logos, Berlin (Mündliche Kommunikation)

20. Petrushin V (1999) Emotion in speech: recognition and application to call centers. In: Artificial neural nets in engineering (ANNIE'99), p 7–14
21. Pfister B, Kaufmann T (2008) Sprachverarbeitung. Springer, Berlin
22. Pittermann J, Pittermann A (2006): Integrating emotion recognition into an adaptive spoken language dialogue system. In: 2006 2nd IET international conference on intelligent environments, IE 06, p 197–202
23. Scherer K (2013) Vocal markers of emotion: comparing induction and acting elicitation. Comput Speech Lang 27(1):40–58
24. Schuller B, Batliner A (2014) Computational paralinguistics: emotion, affect and personality in speech and language processing. Wiley, New York
25. Schuller B, Steidl S, Batliner A, et al. (2012) The INTERSPEECH 2012 speaker trait challenge. In: Proceedings INTERSPEECH
26. Schuller B, Steidl S, Batliner A, et al. (2010) The INTERSPEECH 2010 paralinguistic challenge. In: Proceedings INTERSPEECH, p 2795–2798
27. Stahlknecht P, Hasenkamp U (1999) Einführung in die Wirtschaftsinformatik, 9. Aufl. Springer, Berlin
28. Thompson W, Balkwill L (2006) Decoding speech prosody in five languages. Semiotica 158:407–424
29. Vidrascu L, Devillers L (2007) Five emotion classes detection in real-world call center data: the use of various types of paralinguistic features. In: Proceedings of the international workshop on paralinguistic speech—between models and data. Citeseer, p 11–16
30. Walther M (2018) Automatische Erkennung paralinguistischer Merkmale zur Bewertung der Gesprächsqualität in Callcentern: Zweistufige maschinelle Klassifikation mittels multipler Lernverfahren und perzeptiver Kriterien. In: Hoffmann R (Ed) Studientexte zur Sprachkommunikation, Bd. 89. TUDpress, Dresden
31. Walther M, Neuber B, Jokisch O, Mellouli T (2015) Towards a conversational expert system for rhetorical and vocal quality assessment in call center talks. SlaTE 2015:29–34
32. Walther M, Mellouli T (2017) Intelligente Systeme zur Bewertung der Gesprächsqualität im Callcenter—Stand der Forschung und experimentelle Ergebnisse. In: Neuber B, Pietschmann J (eds) Dialogoptimierung in der Telekommunikation. Bd. 9. Schriften zur Sprechwissenschaft und Phonetik. Frank & Timme, Berlin
33. Walther M, Mellouli T, Jokisch O (2015) Fusion von Klassifikationsmodellen zur automatischen Erkennung von Stimmeigenschaften in der Qualitätsbewertung von Callcentergesprächen. In: Wirsching G (ed) ESSV 2015. TUDpress, Dresden, pp 188–195
34. Weninger F, Wöllmer M, Schuller B (2014) Emotion recognition in naturalistic speech and language—a survey. In: Konar A, Chakraborty A (eds) Emotion recognition: a pattern analysis approach. Wiley, New Jersey, pp 237–268
35. Yacoub S, Simske S, Lin X, Burns J (2003) Recognition of emotions in interactive voice response systems. In: Proc. 8th European conference on speech communication and technology (Eurospeech 2003), p 729–732

Prof. Walther studied Business Informatics at Martin Luther University Halle-Wittenberg and at the University of Wollongong, United Arab Emirates. He began his professional career as a research assistant at the Institute of Information Systems at Martin Luther University Halle-Wittenberg. Subsequently, he worked as an IT analyst and training manager for software architecture. He then worked as a data scientist for an energy supply company. Prof. Walther was also the founder of an Internet travel portal and a lecturer at two universities. His research focuses on mobile applications and pattern recognition in speech data as well as data engineering and data science.

Index

Printed in the United States
by Baker & Taylor Publisher Services